DIGGING GOLD
AMONG THE ROCKIES
Or

Exciting Adventures of Wild Camp Life In
Leadville, Black Hills and the Gunnison Country

BY G. THOMAS INGHAM

U. S. Deputy Mineral Surveyor for the lately discovered
extensive mineral belt of Colorado, Dakota, Montana, etc.

Western Reflections Publishing Company
Lake City, Colorado

A Reprint Published by
Western Reflections Publishing Co.
P. O. Box 1149
951 N. Highway 149
Lake City, Colorado 81235

www.westernreflectionspub.com
westref@montrose.net

Copyright Western Reflections Publishing Company 2008
Printed in the United States
Library of Congress Number: 2008930936

ISBN 978-1-932738-68-1

DIGGING GOLD
AMONG THE ROCKIES

PROLOGUE

Unfortunately, very little is known about George Thomas Ingham, who wrote *Digging Gold Among the Rockies*. We know that he was a U.S. Deputy Mineral surveyor who traveled extensively in Colorado, Dakota, and Montana in the early 1880s. And, some tantalizing glimpses of his life can be found throughout this book. Ingham discusses many subjects that are very important in mining history and especially for the mining camps in Gunnison, Leadville, and the Black Hills during their formatives stages. His book is not limited to just mining. He also writes of gambling and drinking problems, stock swindling, road agents, hunting and trout fishing and much more. He gives practical and realistic advice to the prospective gold-seekers, warning that "all is not gold that glitters." He was an artist who profusely illustrated this book, which is now considered one of the primary sources for researchers working in those areas.

Ingham wrote that "it is not the object of this book to add anything essentially new to scientific knowledge. And it was not written to bear criticism as such. The author has endeavored to place before the public, in language which all can understand, such facts and history as has come to his knowledge in regard to this wonderful and growing industry." Such a goal makes it easy for the non-historian or those who have no knowledge of mining to read and understand this work.

This book is very rare and originals of this work are being lost daily because of the highly acidic paper on which they were printed. Western Reflections Publishing Company is pleased to preserve *Digging Gold Among the Rockies* and make it more widely available to anyone interested in Colorado or mining history.

A PEEP INTO COLORADO.

CONTENTS.

CHAPTER I.

The discovery of gold in the United States—Gold in North Carolina: the earliest discoveries made—Gold in South Carolina, Virginia, Georgia, Alabama and Tennessee—The discovery in California: by whom made—Ancient discoveries in Arizona, 1748—The California Indian—A sharp Yankee trader—How the discovery came to be made in California—A short biography of the discoverer—The mill-race where gold was first found—The test that decided it to be gold—Discoveries in other Territories—Discovery of the Comstock Lode: by whom made—General Custer finds gold in the Black Hills—Professor Jenney's expedition to the Black Hills in 1875—Product of the Black Hills—Annual product of the United States in 1879—Great progress in railroad building to the mining regions—Gold and silver in Arkansas and Missouri—Silver mines in Maine: who discovered them, and how—The cost of living in the early excitement in California—High prices in 1849—The cost of a single meal, $21.50.. 17

CHAPTER II.

Placer mining—Ancient river-beds—Table Mountain, California—Surface mining and deep gravel mining—The blue lead—Quartz the mother of gold—How placer deposits were formed—The bed-rock—Gold-saving apparatus—The specific gravity of gold—Of other metals—Winnowing gold—The pan process—The cradle, or rocker—The Long Tom—Amalgamation with quicksilver—Hydraulic mining—The novel invention, and where first used—The inventor unknown—The sluice—Water brought over fifty miles to dry diggings—Bank blasts of fifty tons of powder—Cleaning up the sluices—Retorting the amalgam—Casting the gold bricks—Rich strikes—A two thousand eight hundred dollar nugget—The largest piece of gold ($6,000)—An eight thousand dollar find—A building torn down for the gold beneath it—Alder Gulch, Montana—Confederate Gulch, Montana—Marvelous diggings—Aladdin-like wealth—One thousand dollars to the pan of earth—Coarse gold and nuggets—Diamonds in California—Diamonds—How to recognize them... 48

CHAPTER III.

Quartz mining—Veins of ore—The Mexican arrastra—The geology of mining—Granite: of what composed: description—Metamorphic rock—Origin of minerals—Where to look for veins of ore—"Fools' gold"—Iron pyrites—Quartz veins: their principal features—The great "Mother Lode" of California—The Comstock Lode: its size and production—The Sutro Tunnel: an address by Adolph Sutro, its projector—Formation of the fissure: how it was filled—The theory of volatilization—Great depth of the Comstock Lode—Bonanzas—Carbonate ores of Colorado—The first quartz-mill—The stamp-mill................................ 95

CHAPTER IV.

Mining laws—How to locate mineral claims............................ 133

CHAPTER V.

The Black Hills: early history and discovery—Indian traditions—Origin of the discovery of gold—The first party to winter there—Evidences of former occupation—The first city—The first discovery of quartz—Cost of some of the mines—Annual production—Mining regulations, etc................................ 139

CHAPTER VI.

Westward ho!—Notes by the way, etc......................... 155

CHAPTER VII.

Black Hills—Division into counties—Streams—Rivers—Creeks—Scenery—Mountains—Peaks—French Creek—Spring Creek—

CONTENTS.   xi

PAGE.
CHAPTER XIX.

Ruby Camp—The town of Irwin—Along the road—Beautiful wild flowers—Grand scenery—Castle Rocks—The coal region—The falls of Big Ohio Creek—Climbing into the Elk Mountains—Snow along the road in July—The town: elevation, population—Rival towns again—Business—Prices of living—A lake ten thousand feet high—A bit of history—The mines of Ruby Camp—Ruby silver ore—Extraordinary richness—The Ute Indian Reservation—Rich in minerals—Sixty miles into the reservation—Living on venison and game—Forty pounds of trout—Indians not hostile—A terrible adventure with mountain lions—Eight days without food—Gothic City—The town of Crested Buttes... 337

CHAPTER XX.

From Gunnison City to South Arkansas Station—Over Marshall Pass—Recrossing the Rockies—Distance and fare—Fifteen hours in a crowded stage-coach—Incidents of the journey—Walking across the summit—Scenery—Poncha Springs—The town of South Arkansas—Rapid growth—On to Denver—The town of Cleora—The "deserted village"—Entering the Grand Canon of the Arkansas—Massive scenery—Through the "Royal Gorge"—Gigantic walls, two thousand feet high—Canon City—Pueblo—Sights along the way—Pike's Peak—Monument rocks—Safe arrival in Denver 349

CHAPTER XXI.

Leadville, the carbonate camp—Early history—California Gulch in 1860—Former rich yield from the placer mines—The "heavy sand" that troubled the sluice-boxes—Found to be carbonate ore—The first mines located—The first sampling works—The first smelter—A store in June, 1877; by whom established—The first buildings in Leadville—Marvelous growth in 1878—Poor men raised to sudden wealth—High prices of real estate—Rents—Business—Smelters—One million dollars per month—A few of the mines—Value of the ores—The Little Bonanza: Triangle—A few of the men who struck it rich—Five millions in profits from mine speculations—Some of the big bonanzas—The R. E. Lee mine—The Chrysolite—The Morning Star—The Little Chief—The Little Pittsburg—Cost of Living—The great strike for wages................ 362

CHAPTER XXII.

Stock gambling—Mining speculations—Tricks of sharpers, etc.—The assessment laws of California and Nevada—" Freezing out " small stockholders—Comstock management—The Consolidated Virginia and California management—Assessments levied in Nevada in 1879—Dividends of the United States in 1879—The Little Pittsburg stock bubble—Great decline in value—A common swindle—" Wild-cat " mines—Tunnel schemes—Reports of professed experts, etc.—Big profits to promoters in stocking mines, etc.—" Salted 'mines," etc... 383

CHAPTER XXIII.

History of a few mining millionaires: men who made their fortunes in mines—Life of Lieutenant-Governor H. A. W. Tabor, of Colorado; Ex-United-States-Senator Jerome B. Chaffee, of Colorado; George H. Fryer, Esq., of Colorado; Archie Borland, Esq., the owner of some Black Hills bonanzas; John W. Mackay, of California; James C. Flood, Esq., of California; James G. Fair, of Nevada; United States Senator William Sharon, of California—Sandy Bowers, a Comstock character sketch........................ 398

CHAPTER XXIV.

Conclusion—A word of advice—All is not gold that glitters—Amount of funds required for the journey—Railway and stage fare—The contingencies of such a journey—Sickness—A noble act—Prospecting a great lottery—Who draw the prizes—The old-timers of '49—Plenty of room for pluck and energy—How to outfit for prospecting—What to provide—The cost—The burro—A useful animal 429

Rapid Creek—Box Elder Creek—Bear Butte Creek—Whitewood Creek—Spearfish River—Game, etc.................................... 167

CHAPTER VIII.

Deadwood City—Early history—Laying out the town—The first cabin—A big sale of goods—The great fire—Rising from its ruins—Remarkable energy in rebuilding—Deadwood rebuilt—The first post-office—Stamps sold—Money-orders issued—Cost of living in Deadwood—Deadwood market report.. 179

CHAPTER IX.

Black Hills gold mines—The first mines discovered—The first quartz-mill: who built it—The Reno mine—Rich ore—Cement mines—Monthly yield of some mines—Four thousand seven hundred locations in the Hills—Veins of enormous width—"The Great Belt;" Theories about its formation—The Desmet mine: its production—The Deadwood mine: its production—The Highland mine: cost—The Homestake: its cost: production—The Giant and Old Abe: cost—The Rhoderic Dhu mine—Production of some other mines—Desmet mine and mill—The Homestake mine and mill—The Caledonia—The Golden Terra........................... 187

CHAPTER X.

Black Hills silver mines—First discovery—The Florence mine and mill—The Cora mine—The Bald Mountain district—The Rockerville placer mines, etc.. 211

CHAPTER XI.

Taking the bullion away—Road agents.. 219

CHAPTER XII.

Coal, oil, salt and agricultural resources, railroads, etc.................... 228

CONTENTS.

CHAPTER XIII.

Discovery of gold in the Centennial State..................................... 236

CHAPTER XIV.

Onward to the Gunnison Country over the backbone of the Continent... 250

CHAPTER XV.

Pitkin... 288

CHAPTER XVI.

The mines of Pitkin... 301

CHAPTER XVII.

Ohio City—The mines of Ohio Creek—Game and speckled trout—Beaver-dams and houses—A new-made grave—An epitaph—A horrible tragedy at Ohio City—Shooting affray—In camp prospecting—Pitching our tent—Around the camp-fire—Baking our own bread—The "Dutch Oven"—Grand scenery—View of the Uncompahgre range—The "camping-out glory"—Cool nights in the mountains—Struck ore—Dreams of sudden wealth—A fortunate "grub-stake"—Description of mines—Reported carbonate strike—Forest fires—Narrow escape from burning—Miners' cabins destroyed... 314

CHAPTER XVIII.

Gunnison City, the county-seat of Gunnison County—The road from Pitkin to Gunnison—Situation and population of Gunnison—Prices of real estate, lumber, etc.—Rival towns—The railway soon expected—The coal-fields north—Its natural advantages and prospective "boom"... 331

APPENDIX.

Valuable tables, showing the yearly product of the United States from 1848 to 1880—Product of the States and Territories west for 1879—Annual product of lead, silver and gold in the States and Territories west of the Missouri from 1870 to 1880—The world's product of gold and silver—The dividends of the mines of the United States for 1879—List of the dividend-paying mines of the United States—Decision of the Commissioner of the General Land Office in regard to the town-site of Deadwood, Dakota—The mining laws of the United States and regulations thereunder.......................... 438

Definitions of mining terms.. 446

AUTHOR'S PREFACE.

IN view of the recent wonderful discoveries in Colorado and the Black Hills, and the interest they have awakened in precious metal mining, the great revival which is taking place in the quartz mining industry, and the increased bullion production which will result therefrom, the author has seen fit to prepare the present volume, deeming that the time was ripe for its appearance.

When it is considered that in 1876 a section of country known as the Black Hills, was little better than a *"howling wilderness,"* but in 1880 was organized into three of the most populous counties of Dakota, producing six millions of gold and silver bullion annually, and that a single district in Colorado (Leadville), which, in 1878, produced an inconsiderable amount of precious metal, did, in 1879, produce *eleven millions of dollars;* that in one year it increased its production nearly eight millions, and that the State of Colorado was, in 1880, producing at the rate of two millions per month, with a prospect of a steady increase, it would seem that an explanation is given why such a wonderful interest has been excited in mining and mining stocks in the large cities of the East.

We have, in the vast region lying west of the Missouri River, and contiguous to the Rocky Mountains, four States and eight Territories, in which mining is being successfully prosecuted. The entire region is favored with a healthful and invigorating climate, and is rich in mineral resources. Almost everywhere the Indians have been subdued, and order and security to person and property prevail. Many of the dangers and hardships that attend pioneer life are past. And to most of these sections the iron rail has reached or will soon be extended. The difficulties in the way of precious metal mining have grown wonderfully less within a few years. Great improvements have taken place in mining and mill machinery, and in the methods of manipulating the ores.

Rock, which formerly baffled the skill of mill men to work at a profit, is now yielding large returns to the producer. The question of transportation, which to the far-off Territories of Montana, Idaho, New Mexico and Arizona was such a detriment to their prosperity, seems now about to be solved. The Utah and Northern Railroad has been pushed from Ogden up into Idaho and Montana. The Southern Pacific from San Francisco, through Arizona eastward; and now that the Atchison, Topeka and Santa Fe has reached the capital of New Mexico, it would seem that we are indeed upon the eve of a wonderfully prosperous mining season.

Gold and silver, and their discovery and production, are subjects which interest everybody. And, although there are many valuable works on this subject, of a scientific nature, and descriptive of quartz mining, yet there are comparatively *few* of them which are interesting to the general reader, and there is still a vast amount of ignorance in regard to the mode and manner of producing the precious metals.

It is not the object of this book to add anything essentially *new* to *scientific knowledge*. And it was not written to bear criticism as such.

PREFACE.

The author has endeavored to place before the public, in language which *all* can understand, such facts and history as has come to his knowledge in regard to this wonderful and growing industry. If the scientist, or learned professor, or mining engineer should be disappointed in not finding here profound expressions upon favorite theories, or should miss such technical terms as are common in some books of this kind, they will please bear in mind that the *masses* of the people, the readers whom we seek to interest, do not want them. Hence we have sought the most plain and unvarnished language, deeming that the simplest modes of expression were most suitable for our purpose; indeed, parts of the book border rather on romance than on the scientific treatise, and if this volume shall prove of service only to the class named, our object will have been attained.

LIST OF ILLUSTRATIONS.

	PAGE.
FRONTISPIECE—A Peep into Colorado.	
Mining Village in Arizona	23
An Original Land-owner	27
Interior of Copper Mine	39
Washing Gold in the Cradle	60
A Mountain Lake	65
Lake Tahoe, Sierra Nevada Mountains	67
Washing Down the Gold Hills	71
Washing Gold with the Shaker	71
Howland's Improved Disintegrating Riffle	75
Mouth of the Mine	111
Hand-drilling in the Mine	111
Deep Mine Work	119
Overhand Stoping	123
Underhand Stoping	123
Sperry's Wrought Iron Frame Stamp Mills	127
Stone and Ore Crusher	131
Sectional View "Little Giant"	131
Bad Lands Mountain	141
Miners around their Camp-fire	147
Basaltic Columns of the Black Hills Region	169
Game in the Black Hills	177
Ingersoll Rock Drill	197
Ingersoll Rock Drill (in mine)	201
The Howland Pulverizer	205
Excelsior Grinding and Amalgamating Pan	209
Summit of Pike's Peak, Colorado	239
Northern Slope of Uinta Mountains	247
Head-waters of the Colorado, Greene River, Wyoming	253
Winnie's Grotto—A Side Canon	265
Devil's Gate, Georgetown, Colorado	275
Mount of the Holy Cross, Colorado	275
Boulder Canon, Colorado	315
Valley of the Tumichi River	321
Gunnison's Butte	333
Grand Canon of the Arkansas, Colorado	355
Business Street of Leadville	363

CHAPTER I.

THE DISCOVERY OF GOLD IN THE UNITED STATES—GOLD IN NORTH CAROLINA: THE EARLIEST DISCOVERIES MADE—GOLD IN SOUTH CAROLINA, VIRGINIA, GEORGIA, ALABAMA AND TENNESSEE—THE DISCOVERY IN CALIFORNIA: BY WHOM MADE—ANCIENT DISCOVERIES IN ARIZONA, 1748—THE CALIFORNIA INDIAN—A SHARP YANKEE TRADER—HOW THE DISCOVERY CAME TO BE MADE IN CALIFORNIA—A SHORT BIOGRAPHY OF THE DISCOVERER—THE MILL-RACE WHERE GOLD WAS FIRST FOUND—THE TEST THAT DECIDED IT TO BE GOLD—DISCOVERIES IN OTHER TERRITORIES—DISCOVERY OF THE COMSTOCK LODE: BY WHOM MADE—GENERAL CUSTER FINDS GOLD IN THE BLACK HILLS—PROFESSOR JENNEY'S EXPEDITION TO THE BLACK HILLS IN 1875—PRODUCT OF THE BLACK HILLS—ANNUAL PRODUCT OF THE UNITED STATES IN 1879—GREAT PROGRESS IN RAILROAD BUILDING TO THE MINING REGIONS—GOLD AND SILVER IN ARKANSAS AND MISSOURI—SILVER MINES IN MAINE: WHO DISCOVERED THEM, AND HOW—THE COST OF LIVING IN THE EARLY EXCITEMENT IN CALIFORNIA—HIGH PRICES IN 1849—THE COST OF A SINGLE MEAL, $21.50.

The Discovery of Gold in the United States.

THIRTY-TWO years ago the discovery was made which peopled the Pacific Coast. California, then a Mexican colony, subsequently became annexed to the United States by a treaty of peace, ratified on the 30th of May, 1848. It was then a vast territory, sparsely settled by a few Mexicans, less Americans, and some Indians, and was mainly governed or controlled by parish priests. Its only productions were hides and tallow, with which vessels were occasionally laden, and for which large numbers of cattle were slaughtered, their carcasses being mainly wasted. With the discovery made by J. W. Marshall, a new era dawned for California. The importance of his discovery may be realized when we consider that within a little more than one year, or at the close of 1849, United States Commissioner King estimated that fifty-five thousand miners were at work in the gold fields, and that forty millions of dollars in gold dust had been taken, and that the probable yield of the succeeding year (1850) would be *fifty millions.*

Such was the revolution which had taken place in this short time, never equaled by any country. This annual product of California increased until 1853, when it reached sixty-five millions.

Gold in North Carolina.

Up to this time Virginia, North Carolina and Georgia had mainly furnished the gold production of the United States. Gold had been discovered there for many years, but nothing had occurred to excite such a general interest in its pursuit or such a mad rush to the gold fields, as the reports which came from far-off California in 1848–49.

Then it was something of an undertaking to cross the continent; an undertaking requiring months of weary trial and hardship to cross the desert plains; or requiring a sea voyage and journey across the isthmus, of nearly or quite as long, and accompanied by the perils of fever and disease. Many were the poor souls who perished in the undertaking.

To-day it is, with the aid of the palace-coach and sleeper, but the pleasure-trip of a week. And what has made this change? What, indeed, has stretched this band of steel across the continent, unless it be the discoveries made in that golden clime.

The Earliest Discoveries.

It is supposed that the first discoveries of gold were made in North Carolina in 1820, in the vicinity of Mecklenburg. There is an old account of a province visited by De Soto, in his expedition of 1538–40 to a province called "Cofachiqui," whose capital and chief town stood upon a tongue of land between Broad River, of Georgia, and the Savannah River, just opposite the modern district of Abbeville. It is said the Spaniards entered this town and found the country ruled by a beautiful Indian queen, named "Xualla." Here they found trinkets of gold, and hatchets formed from a mixture of gold and copper, and they concluded they had found the long-sought-for precious deposits of gold.

"And so they had," says the historian.* "But it was neither their good fortune nor deserts to find out the precise spot where they could be obtained." "In less than fifteen miles south-east of

---

* Logan, in his History of "Upper Carolina."

the town, on the Carolina side of the river, lay one of the most extraordinary gold deposits in the world."

In 1825 the gold mines of North Carolina were well known, and in 1830 the annual production was about half a million. The principal mines were in the vicinity of Charlotte, where, previous to the California discoveries, a United States mint was established. For a long time subsequent to the California excitement, and during the war of the rebellion, the mines were inactive. But within the last two years a revival has taken place, and with the aid of improved machinery the mines will soon produce again equal to their palmiest days. The total production of the State up to the present time is estimated at twelve millions, and this yield is increased by the mines of South Carolina, a continuation of the same mineral belt, about two millions more.

Gold in Virginia and Georgia.

Gold mining in Virginia began in 1829, and up to the present time about four millions have been extracted. The mines are situated in the range from Fredericksburg to Danville, and have been in many places rich. But, until quite recently, work on the mines has been for years abandoned. Here, as in North Carolina, a revival is taking place in the industry, and new quartz mills are being erected.

Georgia, Alabama and Tennessee each have a portion of the same gold-bearing range, and these three States have produced up to the present time at least twenty millions. Georgia has been the greatest producer of any of the Southern States, the mines having been the most actively worked, and at the present time she is producing *annually* nearly half a million. At Dahlonega, this State, many years ago, a United States mint was established, which at the present time is nothing more than a Government assay office. The States above mentioned, previous to the advent of California, were the source of most the entire production of the United States.

The Discovery in California.

The credit of the discovery in California has been universally conceded to James W. Marshall. But it appears there were other discoveries of the precious metal in that State, and undoubtedly its existence was known to the Mexicans much earlier than Marshall's discovery. But it remained for him to make the grand discovery of the richness of those deposits, and to make known the fact which caused the stampede that populated the State.

Ancient Discoveries in Arizona.

Richard J. Hinton, in his excellent book on Arizona, says that, "Although there are indications that mining operations were carried on here (Arizona) in the seventeenth century, it was not until 1748 that the records became definite." "In that year the San Pedro Gold Mine (in Arizona), it is known, was worked by the Spaniards."

Certain it is that, according to an old Spanish map of Arizona in existence, bearing date of 1775, on which some mining villages are located, and according to Spanish records kept, that as many as two hundred silver mines were being worked at that early date. Therefore it is not improbable that Spanish-Mexican miners had extended their search for gold over into California long previous to the time of J. W. Marshall.

In 1842, it is said, a Mr. Dana, who traveled through Upper California, speaks in his work of the "Auriferous Quartz" of the Sacramento Basin, and remarked its resemblance to other gold districts. Thomas O. Larkin, United States Consul at Monterey, in 1816, also wrote to the Secretary of State at Washington, that "at San Fernando" (now Los Angeles), California, "from one to five dollars per day" in gold could be obtained by washing certain black sand. But added that few had the patience to look for it. So it appears that, although known to the Mexicans, it had not been found in quantities to encourage pursuit for

The Lucky Man.

it. Therefore, we consider Marshall entitled to all the honors of a first discoverer. It has also been

would back out of the bargain, suddenly seized the paper of raisins and ran for the woods; the raisins costing the trader about five cents, and the gold amounting to over thirty dollars.

How the Discovery came to be Made.

James Wilson Marshall* was born in Hope Township, Huntingdon County, New Jersey, in 1812. At the age of twenty-one he started westward, stopping first a few months in Indiana, then halting for a time in Warsaw, Ill., and from thence to Missouri, near Fort Leavenworth. Here he had some idea of settling; but after struggling with the fever and ague for a few years, he started, about May 1st, 1844, in company with others, overland to California. They took a northern route, via Oregon, and thence southward; they reached the Sacramento Valley, about forty miles from the present city of that name; and finally went to Sacramento, then called Sutter's Fort, where Marshall engaged to work for Colonel Sutter, who owned a large estate there and trading post, and who had built a small fort, which bore his name.

The Mill-Race where it was Found.

Some time in the year 1847, Marshall was sent by Sutter up the American River, in search of a site for a saw-mill; and going up the south fork

---

* See his biography, by John F. Parsons, of Sacramento, Cal.

asserted that the Indians were the first discoverers, and it is a tradition that, at a very early period after the organization of Catholic Missions in Lower California, that the Indians, who were sent into the upper country to persuade the natives to submit to the guardianship of the Catholic fathers, on their return spoke of the "*shining sand*" in the streams which they had crossed in their journey. But it seems that the story of the shining sand was unheeded. It appears, in other accounts, that the priests rather discouraged the search for gold as demoralizing and injurious to their mission work, which is probably true.

It is very certain that the California Indian, although he may have seen the shining nuggets beneath his feet, never knew their value, and never saved them for any useful purpose. The reason, perhaps, for supposing the Indians to be the discoverers has been because, after the discovery by white men, they learned to mine it, and in some instances became very expert in finding it. A curious incident is related of a Yankee trader, who had set up a small shop among miners, whence he dispensed groceries, tobacco, etc., in the early days of California gold mining. One day an Indian came to the tent with a handful of gold dust, wrapped in a cloth, and putting it in one side of the scales, the trader put raisins into the other, until they balanced, and displayed so little haste in the operation that the Indian, fearing he

AN ORIGINAL LAND-OWNER.

of that river he at length reached a point, afterward known as Coloma, where now stands the town of that name. Having found here a suitable location, and marked a site for a mill, he returned to Sutter's Fort, and formed a copartnership with Sutter, who agreed to assist him in building the mill and in carrying on the lumbering business. This contract was entered into about the 19th of August, 1847. Accordingly work was at once begun, and progressed rapidly, and with the aid of two or three men and some eight or ten Indians, a mill-race was in course of construction. The method employed was in shoveling out a ditch by day, and by turning in the water to wash it out deeper at night. "On the morning of the 19th of January, 1848," says Marshall's biographer,[*] "Marshall went out as usual to superintend the men, and after closing the forebay gate, and thus shutting off the water, walked down the tail-race to see what sand and gravel had been removed during the night. This had been customary with him for some time, for he had previously entertained the idea that there might be minerals in the mountains, and had expressed it to Sutter, who, however, only laughed at him. On this occasion, having strolled to the lower end of the race, he stood for a moment examining the mass of debris that had been washed down, and at this juncture his eye caught the glitter of something

---

[*] John F. Parsons, Sacramento.

that lay lodged in a crevice, or ripple of soft granite, some six inches under the water. His first act was to stoop and pick up the substance. It was heavy, of a peculiar color, and unlike anything he had seen in the stream before. For a few minutes he stood with it in his hands, reflecting, and endeavoring to recall all that he had heard or read concerning the various minerals. After a close examination he became satisfied that what he held in his hands must be one of three substances—mica, sulphuret of copper or *gold*. The weight assured him it was not mica. Could it be sulphuret of copper? He remembered that this mineral is *brittle*, and that gold is *malleable;* and as this thought passed through his mind he turned about, placed the specimen upon a flat stone, and proceeded to test it, by striking it with another. The substance did not crack or flake off; it simply *bent* under his blows. This, then, was *gold*, and in this manner was the first gold found in California."

Another account says Marshall sent the specimen to Mrs. Peter L. Wimmer, who lived at the new mill, and as she had formerly lived at the gold mines in Georgia, he sent it for her opinion. She was boiling soap, and put the metal into the kettle to see if it would corrode, and thus proved it to be gold.

After the discovery, it seems Marshall proceeded with his work as usual, and having quietly collected some three ounces of the metal, he had occasion, about four days afterward, to go down to Sutter's. Taking his specimens with him, he started. Arriving at Sutter's store, he concluded his other business, and then asked for a private conversation with Colonel Sutter. They repaired to a little private office back of the store, where Marshall showed him the gold. He looked at it in astonishment, and asked what it was. Marshall replied that it was gold. "Impossible!" was the ejaculation of Sutter. Marshall asked for some nitric acid to test it, and having sent to the gunsmith's for it, Sutter inquired if there was no other way in which it could be tested, when it occurred to them to weigh it. Accordingly, some silver coin and a pair of balances were brought in, and Marshall proceeded to weigh it. As he expected, the gold went down and the coin of equal bulk rose lightly up. Subsequently the acid having arrived, that settled the question, and all doubts vanished.

A few days afterward, Sutter, Marshall and a man named Humphrey, formed a mining copartnership. In the month following, Marshall and Peter L. Wimmer went to "Mormon Island," and found gold there. Other parties, residing in different portions of California, hearing of the gold

Discoveries in Other Territories.

discoveries at Coloma (Marshall's Mill), came there and observed the indications, and returning home, prospected their own neighborhoods with success. Then the excitement began to spread, but it was many months before the news reached the East. Vessels were not frequent from Atlantic ports.

The Panama Railroad was not then built, and the journey across the plains was a slow and tedious one—a route not frequently traveled. It was late in the year 1848, and not till the beginning of '49, that the rush began which was to revolutionize the Pacific Coast. Since then there have been many new discoveries of scarcely less importance, and many *"rushes"* of a similar character. Gradually as the richest bars and shallower gravel deposits were worked out in California, the miners extended their explorations to other territories, with more or less success. As the vast army of gold-hunters increased, they sought less crowded fields and more secluded haunts, in which to dig for the precious stuff.

In 1852 mining was fairly begun in Oregon, which has been prosecuted with success to the present time. Washington Territory and British Columbia were visited by the hardy miners soon after that date, and gold discovered there. Idaho and Montana were visited in 1860, and rich placer deposits found. From Alder Gulch alone, in Montana, near Virginia City, forty millions were extracted within two or three years, and seventy millions have been taken out of it altogether.

The Comstock Lode, in Nevada, was discovered

The Discoverers of the Comstock Lode.

in 1857, by Allen and Hosea Grosch. Hosea died soon after from a pick wound in the foot, and Allen leaving his record of claims with a man by the name of Comstock, who kept a grocery at Carson, went to California. Allen also died from the effects of exposure on the way over the Sierra Nevadas, and Comstock having possession of the claims, exercised ownership over them, thus giving his name to the greatest lode in the world. Thus began the settlement of Nevada, then a part of Utah Territory. In May, 1859, the Pike's Peak discoveries, in Colorado, were announced. Colorado was then a part of Kansas and Utah Territory, and the present boundaries were not known.

To illustrate what a wonderful civilizer gold is: In twenty-one years Colorado, an unorganized territory lying on both sides of the Rockies, with a population of less than one thousand souls, has become a populous State, with thirty counties, with over two hundred towns and cities, and a population of two hundred thousand, and a taxable wealth of over one hundred millions! What, indeed, but her precious mineral deposits could have produced such a change?

In the years succeeding 1860, and during the American conflict, mining excitements seemed to gradually die out. The great appreciation in prices of labor and supplies, the great cost of transportation, the scarcity of help to work the mines, all acted to discourage precious metal

General Custer Discovers Gold in the Black Hills.

mining, and to make it less profitable. For ten years these causes greatly paralyzed the industry.

In 1874 mining excitements had apparently come to a stand-still. Gold mining had run into a slower, more regular and legitimate business. Capital had apparently absorbed the gold and silver fields, and was quietly grinding out the quartz ore, and smelting it into bullion.

Placer mining was a thing of the past, except as carried on by the hydraulic process, operated by large companies. The miners of 1849 and '52, except those working for wages under the direction of capital, had settled down to other occupations in quiet homes—some here, some there, all over the land. Quiet reigned among the goldseekers. Perhaps many, like one of old, had folded their arms and sighed that there were, so to speak, "no more worlds to conquer," no more gold-fields to subdue.

But the west winds blew down the hillsides and mountain passes ominous sounds! A breath —a whisper at first—then louder—"Custer's soldiers have found *gold!*" "They have found hills of gold!" The Black Hills! "The streams are filled with gold. There are hostile Indians. But there is gold."

Custer's expedition of 1874 returned from the Black Hills. The breeze stiffened to a *gale*. There was gold—they had found it—they had brought it back with them. Surely, there was gold now." Such was the talk on the frontier among a few miners in 1874.

Professor Jenney's Expedition to Black Hills.

Still, some doubted, and in 1875 a scientific party, headed by Professor Jenney, were sent out by the Interior Department, with time and means to make a thorough exploration, and troops to protect them from Indians. With them went *lots of miners*. The expedition returned. There was gold, and no mistake about it. But—the country was an Indian reservation, ceded to them by Government contract, and miners could not be allowed to go there. So said the Government, and the President issued a proclamation warning all persons not to trespass upon the Indians' domain. United States soldiers were sent to bring out from the Hills every miner they could find. But, notwithstanding these facts, many went, and a few, dodging the soldiers and escaping the Indians, passed the winter of 1875-'76 in the Black Hills. Some suffered death from Indians; some nearly suffered starvation, living upon game which they killed for subsistence, which fortunately was at that time abundant. Finally, the Government made a new treaty with the Indians—the Red Man had to make another surrender of territory, and some time in 1876 the Government owned the country, and miners could go there without restraint. Then the *rush* began in earnest, and at the present writing (1880) the population of the Black Hills is estimated at twenty-five thou-

Product of the Black Hills.

sand. Scores of quartz mills have been erected; twelve hundred stamps are crushing ore; the largest quartz mills in the world are there situated, and over six millions of gold are being annually produced.

These gold excitements have usually resulted unfortunately to the many, and only remunerative to the few. They have generally enriched only the capitalist, who came with the necessary means to develop the mines. On the whole, they have resulted in permanent good to their respective localities, and have hastened civilization and settlement of our far west, greatly in advance of what could possibly have been accomplished under different circumstances. That in the end they have been the means of adding great wealth to the nation, and have developed the vast resources of the country—agricultural as well as mineral—which would have otherwise lain dormant, and that they were a powerful agent in building up cities and towns, railroads, mills and machinery, no one can doubt.

Perhaps the last, and by no means the least of these excitements of importance, is the late *rush* to Leadville, Colorado. Here has been built within two years a large city of thirty thousand inhabitants. An amount of silver has been produced in the district never before equaled for the length of time, by any country. A chapter will be devoted to this wonderful camp hereafter.

The Annual Product of the United States in 1879.

Thus gradually it became known that in every State and Territory west of the one hundredth meridian (west from Greenwich), Kansas and Nebraska excepted, that deposits of the precious metals existed.

New Mexico and Arizona mines became producers in 1860, although worked by the Spaniards and Mexicans nearly a century before. But not until the close of our civil war, in 1865, did work upon them become active, and not until the present time has their production been very great. They are now assuming considerable importance.

Therefore, in four States and eight Territories, we have deposits of the precious metals rivaling the richest known mines of the world, and the United States to-day ranks first in the production of gold and silver.

It is estimated from reliable sources that the

total annual product of the United States in gold and silver bullion, for the year 1879, is not far from seventy-four millions of dollars. And there is little doubt but it will reach nearly one hundred millions in 1880.

Owing to the adoption of a new Constitution in California, imposing great taxes and restrictions upon corporations, labor difficulties and other causes, the California production was smaller than usual. Nevada was also below her usual production. But revivals are taking place in each State, and in other sections the increase in production being great, our estimate will undoubtedly be reached.

Not alone in the precious metals has the richness of our country west of the meridian named been proven, but lead, copper, mica and coal have been found in most, if not all of the States and Territories named, in quantities practically inexhaustible. Arizona has probably the richest copper mines in the world, and coal deposits of great extent, which, owing to its isolated and far-off position, could not heretofore be made available. The Southern Pacific Railroad, now completed nearly through the Territory, will soon bring her mines into prominence.

When it is considered that the deep gravel deposits and ancient river-beds of California are far from being exhausted, being, in fact, only partially explored as yet, and that the gold area of this

INTERIOR OF COPPER MINE.

State alone is equal in size to the State of New York, can it be wondered that our Western Territories, with such vast and inexhaustible resources, have suddenly grown into populous States, that small towns have in a short time grown into large cities, and that poor men have, in thousands of instances, become suddenly enriched from the sources named, and that in the number and wealth of their inhabitants, the States of the Pacific Coast are in a fair way to rival those of the Atlantic?

Gold and Silver in Arkansas—in Maine.

We should not, however, overlook discoveries of importance which have been made further east. Arkansas and Missouri have a silver-bearing mineral belt, which promises in the near future to develop considerable richness.

The Mount Ida mining district, in Montgomery County, Arkansas, in the vicinity of the famous Hot Springs, has several mining companies at work. The ore is a gold and silver free milling ore, which gives assays of from twelve to twenty dollars per ton near the surface. Near Silver City, Ark., Galena ore is found, containing silver and lead.

There is a great deal of interest being taken in this new mineral region, and the developments being made in several of the mines, give assurance of a mining district of great promise.

New England is also coming to the front as a silver-producing region.

Maine has long been known to contain small deposits of silver and copper ores, but no interest seems to have been taken in them until about two years ago (in 1878). Blue Hill, Maine, had for some time several companies working copper mines, carrying a small amount of silver. These discoveries at Blue Hill led a young man, by the name of A. A. Messer, who had mined on the Pacific Coast, to prospect for mineral along the sea coast, near the village of *West Sullivan*, about twenty-five miles north-east of the Blue Hill copper mines, in Hancock County. In his walks along the sea-shore, on the spot where is now the Sullivan shaft, he found a ledge cropping out, which he at once felt contained the precious metal. He caused assays to be made, which proved satisfactory enough to justify further examination,

Silver Mines in Maine.

and, in connection with others, without capital, blasting was commenced at the very edge of the water, almost between high and low tide.

A lease from the owners of the land for all of their mineral right for a term of ninety-nine years was obtained for a merely nominal sum. They called their prospect the Sullivan Mine; continued their work sinking on the ledge. Developments were satisfactory. Attention was attracted to the property, and it was purchased by Boston capitalists at a fabulous price for the little town of Sullivan, or, indeed, any town outside of large cities.

A company was organized, with a capital of four hundred thousand dollars, divided into shares of ten dollars each. A mining engineer of Boston was placed in charge, with Mr. Messer, the discoverer, as foreman. They sunk a shaft, following the course of the vein, which inclines at an angle of about forty degrees toward the shore, and commencing with the vein, only twenty-two inches wide at the surface, and following it down to about one hundred and forty feet, it gradually widened out to about seven feet of good paying ore. The fissure is remarkably distinct, and the clay selvages are perfect. A lot of eight tons of ore, sent to Balbach & Sons, Jersey City, for treatment, gave them three hundred and forty-five dollars pure silver from the lot. Balbach & Sons informed them that if the rock had been properly "cobbed," as it should have been, they would have had in that sum the product of about four tons of ore, or about eighty-six dollars per ton. Later developments have increased the richness of the ore.

Who Discovered Them, and How.

This discovery at Sullivan proved to be an incentive to prospecting all over the State. At Cherryfield gold veins were discovered. At Gouldsboro, ten miles from Sullivan, rock was discovered assaying near the surface at one hundred and seventy dollars per ton. Extensions to the Sullivan mine were found, and two or three more companies organized to work on the vein. Thus the "boom" in mining has been inaugurated in Maine, which may prove the beginning of an important industry. Already there are over sixty incorporated companies, and the stock of the Sullivan mine is selling at fourteen dollars per share, of the par value of only ten dollars. So, it would seem that capitalists have confidence in the mining resources of Maine.

The Cost of Living in the Early Days of California.

To give the reader an idea of the great cost of living in the early days of California, when provisions, clothing and tools had to be carted over poor roads, long distances, or "packed" on the backs of mules, over a mere trail for hundreds of miles; when a common Irish potato sold for a dollar, and when a pinch of gold dust just paid for a "chaw of tobacco;" when a doctor did not look at a patient for less than twenty dollars, and bread was two dollars per pound. Such facts we think may help us to an appreciation of the trials which those pioneers of California endured. What was true of California, in this respect, was also true, to a greater or less extent, of all the sections in the far West where mining has been prosecuted.

An extract from the diary of the parish priest

of Sonora, California, in 1852, says :* "Eggs were worth one dollar each; bread, two dollars per pound; chickens, ten to twelve dollars each: a turkey sold for *twenty dollars*, and everything else in proportion."

High Prices of Living in 1849—The Cost of a Meal.

A gentleman writing east from California, in 1849, thus gives the items of cost of a single meal eaten by himself and companion, at Coloma or at Marshall's Mill, the place where gold was first discovered. After calling for the articles at a grocery, he says: "We ate and drank with great gusto, and when we had concluded our repast, called for the bill. It was such a curiosity in the annals of a retail grocery business that I preserve it, and here are the items; it may remind some of Falstaff's famous bill for bread and sack.

One box of Sardines,	$16 00
One pound of Hard Bread,	2 00
One pound of Butter,	6 00
A half pound of Cheese,	3 00
Two bottles of Ale,	16 00
Total,	$43 00

or twenty-one and a half dollars apiece for a single meal."

---
* See United States Commissioner R. W. Raymond's Mining Statistics.

STARVATION RATES.

In an extract* from one of the books kept at Sutter's store, in 1849, at Sacramento, the year after the discovery of gold, goods are charged at the following prices:

Two White Shirts,	$40 00
One Fine Comb,	6 00
Three pounds Crackers,	3 00
One barrel Mess Pork,	210 00
Two pounds Mackerel,	5 00
Four pounds Nails,	3 00
One paper of Tacks,	3 00
One pair of Socks,	3 00
One pound Gunpowder,	10 00
One Hat,	10 00
One pair of Shoes,	14 00
Thirteen pounds Ham,	27 00
Thirty pounds Sugar,	18 00
One keg of Lard,	70 50
One pair of Blankets,	24 00
One pound Butter,	2 50
Fifty pounds Beans,	25 00
Two hundred pounds Flour,	150 00

Here is an authentic and reliable schedule of prices during the great rush, and it is not to be wondered at that hosts of miners became intensely poor under the burden of such high prices, and suffered every privation, and almost starvation itself, for the sake of searching for gold.

That any one could make anything, can only be accounted for by the proportionately high wages

---

* See John F. Parson's Biography of J. W. Marshall.

paid for all kinds of labor, and the great richness of the gold-fields, which enabled some lucky miners to make rich strikes and proportionately "big pay" from the auriferous sands and placer mines of the State.

In the next chapter will be found a description of Gold-Saving Apparatus, and a record of some of the "rich strikes" of the early days of gold mining.

CHAPTER II.

PLACER MINING—ANCIENT RIVER-BEDS—TABLE MOUNTAIN, CAL.—SURFACE MINING AND DEEP GRAVEL MINING—THE BLUE LEAD—QUARTZ THE MOTHER OF GOLD—HOW PLACER DEPOSITS WERE FORMED—THE BEDROCK—GOLD-SAVING APPARATUS—THE SPECIFIC GRAVITY OF GOLD—OF OTHER METALS—WINNOWING GOLD—THE PAN PROCESS—THE CRADLE, OR ROCKER—THE LONG TOM—AMALGAMATION WITH QUICKSILVER—HYDRAULIC MINING—THE NOVEL INVENTION, AND WHERE FIRST USED—THE INVENTOR UNKNOWN—THE SLUICE—WATER BROUGHT OVER FIFTY MILES TO DRY DIGGINGS—BANK BLASTS OF FIFTY TONS OF POWDER—CLEANING UP THE SLUICES—RETORTING THE AMALGAM—CASTING THE GOLD BRICKS—RICH STRIKES—A TWO THOUSAND EIGHT HUNDRED DOLLAR NUGGET—THE LARGEST PIECE OF GOLD ($6,000)—AN EIGHT THOUSAND DOLLAR FIND—A BUILDING TORN DOWN FOR THE GOLD BENEATH IT—ALDER GULCH, MONTANA—CONFEDERATE GULCH, MONTANA—MARVELOUS DIGGINGS—ALADDIN-LIKE WEALTH—ONE THOUSAND DOLLARS TO THE PAN OF EARTH—COARSE GOLD AND NUGGETS—DIAMONDS IN CALIFORNIA—DIAMONDS—HOW TO RECOGNIZE THEM.

Placer Mining.

GOLD mining is divided into two classes. Placer mining and quartz mining.

Placer mining is understood to constitute all deposits of gold not found in the shape of veins of ore, or incased within walls of rock, but are those deposits of gravel or alluvial soil containing particles of gold which have been, by the forces of nature, loosened from their original quartz matrix and distributed through the gravel beds and sands of the earth, wherever the agency of water could carry them—in a pure state and generally free from admixture with other minerals.

By quartz mining is understood veins of gold or silver found in the solid rock in the shape of ores, either free or mixed with other minerals, or with each other, but which must go through a mill process to become available for any useful purpose.

Placer mining may also be divided into two classes, surface mining and deep gravel mining. The former operated with water upon the surface of the earth by means of "*sluicing*" or other washing processes, the latter by tunnels or shafts, or both, into ancient channels or river-beds, for the purpose of bringing to the surface the auriferous gravels to be washed. The latter also requiring the aid of capital to bring up the riches from below, and the former being the kind of deposits chiefly sought for and which can be worked by individual labor single handed. Hydraulic mining is placer mining worked by the aid of large streams of water thrown against a gravel bank or hill, at a great pressure by means of flumes, ditches and cotton hose, from a point high above the place to be washed, the water being thrown with tremendous force against the earth, removing it very

Ancient River-Beds.

rapidly. Hydraulic mining usually belongs to the class here termed surface mining. But the process has been applied to both kinds of gravel mining in California.

The class of deposits known as the ancient river

channels or the "*blue lead*" of California, and which kind of deposits seem to be confined to that State, are gold-bearing gravels found deep beneath the surface, and are frequently found beneath a mountain of lava, or volcanic earth. They are penetrated by long tunnels or inclined shafts until a stratum of gravel and clay is struck, showing plainly the evidences of an ancient river-bed, which is usually rich in gold.

A man by the name of T. A. Ayers while prospecting near Table Mountain, California, was led to believe that the hill composed of lava, of that name, rising some twelve hundred feet above the surrounding country, had formerly been the course of an ancient river. A tunnel was commenced by him, and after some progress had been attained, he became discouraged and it was abandoned, but others carried it through, and struck the interior bed of an ancient channel, in which were found gold-bearing deposits of fabulous richness. The news spread, and the country was quickly staked off into claims. Other mines of a similar class were found, and the bed of the ancient river has been traced for several miles. Some of these are among the richest mines of California. These require capital to work them successfully, and are usually worked by large companies. The auriferous character of this ancient channel was first discovered in 1854, near Shaw's Flat, in Tuolume County, California, and, it is said, was discovered accidentally by some miners working where the denudation of the lava crust exposed the channel, and the richest ground of this part of the State.

Deep Gravel Mining.

As late as 1855-56 some claims yielded ten to twelve pounds of gold per day, for many consecutive days, or from two thousand to two thousand five hundred dollars per day. In the excitement which followed the discovery almost the whole length of the mountain was located, and hundreds of tunnels run to strike the channel, many of them without success, and resulted in failure.

The Hughes' claim, on the channel, is worked by a tunnel, and pays about five dollars per cubic yard of gravel, which is mined and run out on cars and washed in sluices. The width of the channel varies from one hundred and fifty to three hundred feet, and the gravel is about twenty feet thick, but the paying portion is about four feet of the bottom. Other claims have developed pay of about the same description, but some were immensely rich.

The Turner's flat gravel deposit, a part of the Table Mountain channel, yielded about one million dollars before operations were suspended; the gravel at the bottom having paid as high as five dollars per bucketful; ten dollars per cubic yard has been frequently averaged for a time in these gravels.

Very few of the miners of 1849-50 knew anything about the scientific principles of gold and silver mining, as now carried on. Perhaps had they known more of the geological structure of the rocks, much needless prospecting would not have been done. But gold was so plentiful then, that search for it in the gravel along the streams, in the bars, and in the bottoms of the creeks and rivers, was all that was attempted, without considering the source from which came all these deposits.

Quartz, the "Mother of Gold."

Hence it was that search for gold was instituted almost everywhere. The soil was turned up for that purpose in many places entirely barren of auriferous deposits; and we presume that but a very small portion of the miners knew but what gold *grew in the ground*, and was originated in the *shape they found it*—in the form of "dust," and that such was its natural state. They undoubtedly learned better by experience a little later; but it is well known that the early miners paid very little attention to quartz veins. Such veins were passed by as of little account in those early days. They had not learned that gold in its original state as a mineral was imbedded in veins of quartz usually within granite, slate or porphyry walls, and was not scattered along the streams, in banks of gravel, in the form of gold dust—except as it had been washed out of the rocks by the streams, and worn down from the ledges by the elements—the winds and the rains—and had been carried down by the floods from the mountains, and deposited in the sands along the valleys. Such was the origin of placer mines, and therefore quartz has been termed the "Mother of Gold."

Later, quartz prospectors learning these facts, used them to their advantage in searching for veins of quartz or lodes of ore, and following up the drift or "float" rock, which had been carried down by the streams, found the original veins from which they came.

How Placer Deposits were Formed.

Such is still the manner of searching for veins of ore. As to the origin of placer deposits and deep gravel beds, or ancient rivers of California and other sections, it should first be understood that the accumulation of these deposits was a very *slow* one, and humanly speaking, a *very long* one, and probably occupied hundreds of thousands of years.

At the time their formation began, the crust or surface of the earth was everywhere a solid rock, uncovered by any alluvial deposits, or soil, or vegetable growth of any sort. The naked rock, barren and rugged, and diversified as now by mountains, valleys, hills and depressions in its surface, was all that could be seen upon the earth. Then there came a time when rains and snows began to fall upon the mountains, and water began to flow down the hillsides, and follow the depressions of the rocky surface. Springs began to issue from the crevices of the rocks, creeks and rivers began to seek the lowest valleys and passes between the hills and mountains— small ones at first, because rain and snow were not plentiful at once, but storms gradually increased in size and extent. Small streams gradually became mountain torrents, cutting into their rocky beds deep channels, and at other points filling up valleys and depressions; ponds and lakes of still water, with the debris they washed from above. Frost came to crack and loosen particles of rock from the overhanging walls of mountains, winds and hurricanes blew down dust and particles of stone; all these agencies, working together to produce an alluvial deposit of sand, gravel and earth, which always lodged in the lowest valleys and sheltered depressions of the then nude surface of the earth, gradually filling them up to a level with higher ground.

The streams in whose channels the ancient gravels were accumulated, were probably shallow ones, and were subject to great variation in the amount of water they carried. Sometimes they were but small rivulets, at others mountain torrents, completely filling up and running over the banks of the depression they occupied, into other depressions of the surface near at hand, carrying their debris with them into others, and forming entirely new beds or courses. Thus, as the valleys they first occupied became filled up or obstructed, they were constantly changing their courses into other and new channels; thus the process went on, slowly but surely.

The time occupied by these changes was a long one—longer perhaps than geologists have estimated. At length, over a vast extent the lower depressions became filled up, and rose higher and higher until every square foot of the bed-rock was covered to a greater or less depth with these deposits, excepting, of course, the highest mountains.

It must have frequently happened, therefore, that large rivers, after occupying one channel for ages perhaps, in the course of time entirely changed their courses. Probably often getting miles away from their original beds, sometimes even crossing the old channel obliquely or at right angles. Thus it may be seen how ancient river bottoms and gravel banks may be found at the present day hundreds of feet beneath the surface, or, as in the case of Table Mountain, California,

The Bed-Rock.

beneath a hill of lava, deposited there by volcanic action after the river-bed was formed.

As will be seen, the gold was originally deposited in cracks or fissures in the naked rocks we have described. These have been properly called veins.

As the streams crossed or traversed these veins of ore, as they must have done, since they swept over, in their changing courses, nearly every foot of the earth's surface, they worked down from the mountains above and from wherever such veins existed the particles of gold and the nuggets loosed from their quartz matrix by the influence of water, the atmosphere and the frosts, and deposited them in the valleys below along the streams and among the gravels we have described. Gold being most the heaviest of all substances, naturally sought the lowest places in the channels and became deposited richest next the bed-rock as found at the present time.

Hence it is that in placer mining miners always search to find the bed-rock, as nearest to the rock and in the waves and uneven depressions of its surface they find the richest pockets of auriferous earth.

The early miners gradually learned by experience some of the facts related above, and began to put them to practical use in determining their search for the precious metals, and to adopt methods and contrivances to aid them in collecting the metal, which they gave various names, such as "The Rocker," "The Long Tom," "The Sluice" and "The Hydraulic," some of which we will describe.

All gold-saving contrivances are constructed in view of the principle of the greater specific gravity of gold over that of all other minerals or substances, except platinum (which is a rare mineral,

Gold-saving Apparatus.

and less plentiful than gold). The specific gravity of platinum being 20.98; of gold, 19.26; silver, 10.5; mercury or cinnabar (quicksilver), 8; copper, 8.5; iron, 4 to 5. The superior gravity of gold, therefore, will cause it to sink even in swift-running water, while the dirt and other substances is being carried on by the current, or as in the winnowing process, described later, allowing the dirt to be blown away in the wind, the gold being retained.

During the first excitement of gold discovery, miners rushed to the scene without tools, and almost without contrivances of any kind. As tools and materials of all kinds were extraordinarily high, they were compelled to use the most ordinary and poorly-constructed machines; in some instances having to hew their timber out of logs and split out their boards by hand, and hew them into shape with axes.

Such being the state of things, they could only use such methods as were within their reach, and it is asserted that one of the earliest, and a favorite method, was simply by the use of a common sheath-knife; to traverse the country and to pry out the shining particles with the knife from the stones and decayed ledges, and in picking up all that could be found on the surface. In this way, it is said, thousands of dollars were gathered during the first year in California.

"Winnowing."

A process called winnowing, or the dry process, was of very early origin among the Mexicans, which they used in diggings that were found distant from streams of water. The auriferous earth was first collected and thoroughly dried in the sun, and reduced to a fine powder. The process then consisted in putting the earth in a blanket, and two men, one taking each end of the blanket, holding it by the corners, commenced throwing it up in the wind, the dust or earthy particles blowing away with the wind, and the gold, by its superior gravity, falling back on the blanket.

This process was continued until only the gold dust and coarser dirt and stones remained, which were separated by washing in the pan.

The Panning Process,

which is still the method used by prospectors in new regions, consists of a tin or sheet-iron pan, not unlike an ordinary milk-pan. Selecting the richest auriferous earth, the operator places a small portion in the pan, water is introduced, when he proceeds to swing it around in a circle, horizontally, by a quick movement, which loosens the dirt from the gold by the action of the water, care

WASHING GOLD WITH THE CRADLE.

The "cradle" was an advance upon the "pan," as a means of separating the gold from the dross. It was a rough box, open at one end and and at the top, and set upon rockers, so that a swaying motion could be imparted to it by the hand. The earth, containing the precious metal is thrown into a box at the top of the cradle, and water is baled up from the stream or pool and poured into the box. The water-flow and rocking motion separate the particles of gold, which are caught by cleets upon the bottom of the machine.

being taken not to upset the dish. The centrifugal force of the movement, and the superior gravity of the gold causing it to settle at the bottom of the dish, a portion of the dirty water being "slopped' o'ver the sides at each turn of the hand. This is persevered in until the water is nearly all thrown out, and only the coarser dirt and gold remains, which can be separated by hand. This operation requires some skill, but miners soon become expert in it. In this way much gold was saved in the days of '49.

"Cradle, or Rocker."

Another contrivance was the so-called cradle, or rocker, which is in some instances still in use in Chinese camps and new districts, consisting of a rude trough, with a plank or board for a bottom and two side pieces. It was from five to twelve feet long, closed at one end, the other end being left open to admit water. Underneath were two rockers, similar to those on a cradle or chair, usually resting on boards, which allow of its being rocked easily. A small stream of water is admitted at the open end, and allowed to run down the inclined bottom toward the closed end, the box soon becomes filled and running over with water at the closed end. The auriferous earth being shoveled into it, and rocking begun, the dirty water is splashed out similar to the panning process, the gold and pebbles settling to the bottom, and most of the dirt escapes with the water. The water can be shut off, the pebbles thrown out, and the process repeated until the gold only remains.

Another style of rocker, much in use, was one of four sides, upon rockers also. It was, perhaps, two feet wide by four feet long, made with two compartments, a smaller box setting on top, and at one end of the larger one, of perhaps two feet square, had either a slat or a wire-sieve bottom, which allowed the gold and fine dirt to be washed through into the box beneath. Into this smaller box the earth was placed, the water dipped into it with one hand, and rocking commenced with the other; the principle of saving the gold being the same as in the rocker first described. The latter machine, however, allowed the water to pass off more freely than the former, having a place of escape in the under box, and also cleats across its bottom to retain the gold.

We shall describe in this connection one more machine, which succeeded the cradle, and was an improvement upon it. It was called the "Long Tom."

The Long Tom

is similar in construction to the longer cradle first described. But it is wider and longer than that, being about twenty feet long, and is shallow. It is placed at an incline of a few degrees from level, and a stream of water introduced into the upper end. About midway in its length an iron or wire sieve turns up on an incline from the bottom, through which the water and fine earth must pass, but which is intended to prevent the pebbles and larger stones from passing further down. Below the sieve are riffles or cleats, nailed across the bottom in several places, against which the gold dust is expected to lodge. Into this running stream of water the gold-bearing earth is shoveled, and is constantly stirred up by one man, and usually a fork with many tines, resembling a manure fork, is used to throw out the gravel and stones which collect on the sieve. In this way several tons of earth can be washed in a day, and at night the stream stopped out, and the gold dust collected from above the cleats, where it has lodged. Sometimes quicksilver is placed above the cleats, to assist in catching the gold by amal-

Amalgamation with Quicksilver.

gamation, a process used in nearly all the hydraulic machines, and which we will describe.

It appears that gold and quicksilver are peculiarly and strongly attracted to each other, and that quicksilver, although not so heavy as gold and silver, is still of a much greater specific gravity than water, and will sink in quite a strong current.

The miners having learned of the great attraction of the two metals for each other, and quicksilver being comparatively cheap (being worth about forty cents per pound), they have used it very largely in both hydraulic and quartz mining operations, and it contributes very much to the success of such mining. In hydraulic mining, which we will proceed to describe, quicksilver performs a very important office, and miners, at this time, would hardly know how to get along without it. The economy of human labor effected by the hydraulic method of mining, over that of all others, is so great that a Mr. Black, mining expert of San Francisco, in a report upon some hydraulic mines of the State of California, in 1864, estimates that, 'While the cost of handling a cubic yard of auriferous gravel with the pan is twenty dollars, with the rocker five dollars, and with the Long Tom one dollar, with the hydraulic process it is only twenty cents." Therefore, the reader can perceive what great improvements have been made over the methods described above. We would here say that the figures in regard to hydraulic mining, being old, are now somewhat behind the times, and that gravel containing twenty cents to the cubic yard has been made to pay good profits to the hydraulic miner, as will be learned hereafter.

Hydraulic Mining

dates back to 1852, a time when the shallower and richer surface deposits began to be worked out. Everywhere along the streams where bed-rock was near the surface, it had been denuded and cleaned of its precious deposits. There still re-

A MOUNTAIN LAKE.

LAKE TAHOE, SIERRA NEVADA MOUNTAINS. (6,220 feet above the sea—21 miles long and 12 miles broad.)

mained hills and mountains, and vast beds of auriferous gravels, where bed-rock was from twenty to hundreds of feet beneath the surface, too deep and the gold too unevenly distributed through them to pay the ordinary pan miner to wash it. Necessity then became the mother of invention. Some new and cheaper method must be devised to remove these deposits and extract their riches.

Thus stood matters in 1852, when a miner, whose name is not now known, put up a novel machine on his claim, at "Yankee Jim," in Placer County, California. He had dug a ditch or race from a small stream, conducting it along the mountain side until opposite his claim, from which point he constructed a flume or spout upon trestles part way across the valley until a "head" or perpendicular height of forty feet was attained. Here he discharged the water into a barrel, at the bottom of which was attached a hose made of rawhide, six inches in diameter, ending in a tin nozzle, tapering down to one inch.

This stream, when directed against his bank of gravel, with the force given by the "head" attained, caused it to wash away and melt before the fierce current with astonishing rapidity. It was simple in its construction, and easily directed to any desired point.

The stream, after playing against the bank, was caught below in a "sluice" or flume of boards many feet in length, through which the muddy water and debris was rushed over cleats or riffles, in which quicksilver was placed for catching the gold.

The news of the invention spread among the miners. Its usefulness was at once apparent. Others were not slow in copying after this ingenious "hydraulic," as it was named, and the foundation of a new system of mining was lain. Improvements became at once in order. Better hose from cotton-duck or canvas was supplied. Nozzles of iron, longer ditches, iron pipes, greater vertical heights were attained, until a power sufficient to remove acres of gravel and hundreds of cubic yards of earth per day was brought into requisition. Thus this little invention, small at first, grew and increased in power until mountains were leveled, and hills were removed and washed down through the sluice-boxes. The whole face of nature was changed by it. The bed-rock became denuded, great boulders uncovered, and the whole country, in course of time, became torn and rent as if with an earthquake.

But there came a time when difficulties arose to check its progress. Water was scarce. Many banks of gravel were situated too far from water or too high above it to be reached.

Most of the claims where water was available were small ones, owned by many different individuals, and too small to pay for the expense of flumes and ditches.

WASHING DOWN THE GOLD HILLS.

WASHING GOLD WITH THE SHAKER.

Cement deposits of gravel were found so hard as to defy even the powerful hydraulic to dissolve them, and the miner of small means found a limit to the profitable working of his new machine. At this time capital came to the rescue. Canals or ditches from twenty-five to fifty miles in length were constructed, supplying plenty of water, which was leased to small claim-holders at so much per cubic inch. Small claims were purchased and consolidated. Bank-blasts of from five to fifty tons of powder were exploded to prepare the ground for the action of water; vertical heights obtained sufficient to reach the dryest placers; chasms of a thousand feet vertical depth were successfully crossed by means of huge iron pipes, carrying water from mountain lakes, and hydraulic mining became a favorite field for the investment of capital.

The foregoing will partially describe hydraulic mining as now carried on. It became necessary in some instances to run long tunnels into the bank or hill, to furnish an outlet for the water and debris, and a grade low enough for the sluice-boxes. These tunnels were generally run in until bed-rock was reached, when a vertical shaft connected the end of the tunnel with the surface. In these tunnels the sluice-boxes are lain—usually they are four to six feet wide, and from one to three feet high, with a grade of six inches in twelve feet, or about four feet per hundred feet. The bottoms of these flumes are generally paved with stones set on edge, or with pieces of plank or joists about six inches in length, also set on their ends, with small cracks between them. This is to prevent the stones and rocks from wearing out the bottom of the flume, and to act as riffles to catch the fine gold. When all is in readiness water is turned into the shaft, down through which it rushes and out through the sluices in the tunnel, gradually tearing down the mountain in its course, and depositing the precious metal in the riffles we have named. After washing has been commenced a day, quicksilver is introduced into the sluice-boxes to catch the gold by amalgamation. The object of the delay of twenty-four hours being so that the interstices between the riffles may get partially filled with gravel and sand before introducing the quicksilver; otherwise it would get too much concealed beneath the pavement to be instrumental in attracting and collecting the gold. The sluice-boxes are charged about twice a day, by means of an iron sprinkling-pot made for the purpose. A light spray of quicksilver is scattered all along the sluices.

After the shaft mentioned has become washed out so large that it no longer resembles a shaft, but is a vast crater, with high sloping gravel banks on all sides, then the hydraulic, with its iron pipes and nozzles, begin to play their fierce currents against them. Sometimes several are in use at

LOWER FALLS OF YELLOWSTONE.

one time, which concentrate their "fire" upon a given point when necessary, and the work of denuding the bed-rock for acres, and often for miles in extent, has begun.

"Retorting."

Once or twice a month, or oftener, according to the richness of the bars or banks being washed, occurs what is called "the clean-up." To do this the riffles or paving-stones are removed, commencing with the upper end of the sluice. A small stream of water is allowed to pass through during this operation, and great care is taken to wash any particles of sand or dirt that might contain fine gold from the blocks of pavement before their removal. The riffles all being removed, and such particles of gravel or dirt as may remain in the boxes being carefully washed and removed, the amalgam, or the gold and quicksilver, is shoveled up into iron or wooded buckets. Scrapers and knives are used to collect any which may cling to the fissures or cracks in the wood. After it has all been collected, it is taken to a furnace, made for the purpose, and when properly washed in a quicksilver bath to separate any lead or foreign substance it may contain, the amalgam, after the surplus quicksilver has been pressed out of it through a canvas cloth, is retorted, and the gold is separated from the quicksilver.

The amalgam is placed in an iron retort, lined

with moist clay; the retort is placed on the furnace, a pipe leads from the retort into an iron

Retorting.

bucket filled with water. The heat causes the quicksilver to pass off in the shape of vapor into the bucket of water, which immediately condenses the vapor again into quicksilver, the gold remaining pure in the retort.

The gold is then placed in a crucible, and cast in a mould in the shape of a brick, is stamped with its weight and name of the company, and is ready for market.

Great care has to be taken in retorting to close hermetically the retort, to allow no vapor to escape, as the vapor from quicksilver is very poisonous if inhaled, and will "salivate" or seriously poison those who may breathe it into their lungs.

Having given a brief description of placer mining and the apparatus by which gold is collected, we will give below some of the earnings from gold washings by means of these machines, and some of the rich strikes which show the wonderful profits which are sometimes made from gold mining. These finds which we have given are of comparatively modern date. Older and much richer ones have been recorded, but these may be considered reliable, and we have taken pains to publish nothing which is not backed by good authority.

In the summer of 1849, a young man named Hudson, from the State of New York, while mining in California, discovered a deep cañon between the town of Coloma and the middle fork of the

Rich Strikes—A $2,800 Nugget.

American River, and by digging some four feet deep reached the granite bed-rock, on which lay immense masses of gold. In the course of six weeks he had dug some twenty thousand dollars from the gulch in gold. The largest piece found in this cañon weighed a little over fourteen pounds of nearly pure metal, and sold for about two thousand eight hundred dollars.

A boy, by the name of John C. Davenport, nineteen years of age, took from this same gulch in one day eleven hundred and fifty-five dollars, and the next day nearly thirteen hundred and fifty dollars of pure gold.

A young man, named Samuel Riper, formerly from Waterloo, New York, with four companions, went on to the Yuba River in June, 1849, and built a dam across about fifty miles above its mouth, thereby laying bare its bottom for a space of seventy feet long by twenty-five feet wide. After a couple of weeks of hard labor, they succeeded in perfectly drying this part of the river's bed, and commenced washing the earth they found in it, consisting of red gravel, solidly packed into the crevices of the rock. The earth proved rich, and yielded about three hundred dollars per day; and in less than two months the party of four divided among themselves the sum of fifteen thousand dollars. Immediately above this two of the same party drained a much smaller portion of the river's bed, and in two weeks took out three thousand dollars' worth of gold.

The Largest Piece of Gold—$5,000.

In August, 1849, Dr. H. Vandyke, with a company of about thirty men, went to the north fork of the American River, and constructed a dam across it just above its junction with the main stream. Within the first three days after drainage was completed the company took out fifteen thousand dollars, and for nearly a month afterward averaged from seventy-five to one hundred and eighty dollars a day per man.

One of the largest pieces of gold ever found, of which there is a record kept, was picked up in a dry ravine near Stanislaus River, California, in September, 1848. It contained a large admixture of quartz, and weighed a little over twenty-five pounds, and was worth five thousand dollars.

A gentleman,[*] who kept a diary while at work in the mines in 1849, which was published in 1850, thus speaks of his success in gold mining on the middle fork of the American River, California:

---

[*] Lieutenant E. G. Buffum, United States Army, " Six Months in Gold Mines," 1850.

"We had packed on the back of one of our mules a sufficient number of boards from Coloma to construct a machine with which to wash for gold. The morning after our arrival we set two of our party at work on the machine, while the rest of us were to dig; and taking our pans, crow-bars and picks, we commenced operations. Our first attempt was to search around the base of a lofty boulder, which weighed probably some twenty tons, in hope of finding a crevice in the rock on which it rested, in which a deposit of gold might have been made. In this we were successful.

"Around the base of the rock was a filling of gravel and clay, which we removed with much labor, when our eyes were gladdened with the sight of gold, strewn all over its surface, and inter-mixed with blackish sand. This we gathered up and washed in our pans, and ere night four of us had dug and washed twenty-six ounces of gold, being about four hundred and sixteen dollars.

* * * * * * *

"The gold which we found the first day was principally procured by washing; although two pieces, one weighing thirteen and the other seven-teen dollars, were taken from a little pocket on the rock. We returned to camp exceedingly elated with our first attempt.

"The next day our machine being ready, we looked for a place to work it, and soon found a little beach, which extended back some five or six yards before it reached the rocks. The upper soil was a light black sand, on the surface of which we could see the particles of gold shining, and could, in fact, gather them up with our fingers. In digging below this we struck a red, stony gravel, that appeared to be perfectly alive with gold, shining and pure.

"We threw off the top earth and commenced our washings with the gravel, which proved so rich, that, excited by curiosity, we weighed the gold extracted from the first washing of fifty panfuls of earth, and found seventy-five dollars to be the result. We made six washings during the day, and placed in our common purse that night a little over two pounds—about four hundred dollars' worth of gold dust.

"After working three days with the machine, the earth we had been washing began to give out, and it became necessary for us to look for a new place. Accordingly we commenced 'prospecting.' I sauntered on ahead of the party, and crossing the river, I continued my search, and after digging some time struck a hard, reddish clay, a few feet from the surface. After two hours' work I succeeded in finding a pocket, out of which I extracted, in addition to other dust, three lumps of pure gold and one small piece mixed with quartz. Elated with my success, I returned to camp, and found the first lot amounted to twelve and a half ounces, or two hundred dollars; and the four lumps last found, sixteen and three-fourth ounces, or over two hundred and fifty dollars. The largest piece weighed no less than seven ounces troy.

"My success this day was, of course, the result of accident, but another of the party had also found a pocket containing about two hundred and seventy dollars, and a place which promised a rich harvest for our machine.

* * * * * * *

"Tired of the old ravines, I started one morning into the hills, with a determination of finding a new place where I could labor without being disturbed with the clang of picks and shovels around me. Striking in an easterly direction, I crossed a number of hills and gorges until I found a little ravine, about thirty feet in length, embosomed amid the low, undulating hills. It attracted my attention, I know not why, and clearing off a place about a yard in length, I struck soil which contained gold. The earth on top was a light black gravel, filled with pebbly stones, which apparently contained no gold. Below this was another gravel, reddish in color, and in which fine particles of gold were so mingled that they shone and sparkled through the whole of it.

"A little pool of water, just below me, afforded a favorable place to test the earth, and scooping up a handful, I washed it, and it yielded about two dollars. I continued digging and washing, until I reached slate-rock, in the crevices of which I found many little nests and clusters of gold, some of them containing eight or ten dollars.

"I flattered myself that I had here at last found a quiet place where I could labor alone and undisturbed, and appropriate to myself the entire riches of the whole ravine. When I reached and had explored the surface of the slate-rock, I tried the experiment of breaking the rock itself into small pieces and washing it; this proved as rich as the gravel, turning out two dollars to the panful. The results of that day's labor were one hundred and ninety dollars' worth of gold dust, and I returned to the house with a most profound secrecy resting on my countenance, and took good care not to expose the good luck I had experienced. But either my eyes betrayed me, or some prying individual had watched me, for the next morning, when busily at work on my ravine, I found myself suddenly surrounded by twenty good, stout fellows, all equipped with their implements of labor.

A $6,000 Nugget.

"I could say or do nothing. Pre-emption rights are things unknown here, and the result of the matter was that in three days the little ravine which I had so fondly hoped would be my own property, was turned completely upside down. About ten thousand dollars was extracted from it, of which I realized a little over a thousand. Merely the body of the ravine, however, was dug, and after it was entirely deserted, many a day I went to it solitary and alone and took from one to three ounces out of its banks."

The same writer also states that, "a few months after the discovery of gold in California, I saw men in whom I placed the utmost confidence, who assured me that for days in succession they had dug over *five hundred dollars* per day."

In September, 1871,[*] a nugget worth six thousand dollars was taken from the claim of Bunker & Co., in the State of Oregon. At Kanka Creek, Nevada County, Cal., in October, 1871, a piece of quartz gold was found weighing ninety-six ounces, and worth fifteen hundred dollars. The same man took from his claim in one day eighteen ounces, worth about two hundred and seventy-five dollars.

Near Carsons, in California, in 1860, a twelve-pound lump of gold was found, slightly mixed with quartz, and valued at two thousand dollars. The lucky miner who found it had been prying out from next the bed-rock a nest of smooth stones, which he scraped clean before throwing them into the heap. One of them struck him as being rather heavy, but it did not occur to him as

"**Rich Strikes.**"

being gold until, in scraping the supposed stone, the yellow metal reflected the rays of his candle.

Some idea of the richness of the California

---

[*] See Report of United States Mining Commissioner, R. W. Raymond.

placers may be learned from an extract taken from the *Auburn* (Cal.) *Stars and Stripes*, a paper of June 15th, 1871:

"From the Weske Claim, twenty men working six days, the yield left to the owner a dividend (or net profit) of four thousand and thirty dollars for the week, or thirty-three dollars and fifty-eight cents per man a day, in gold.

"Last Saturday, John Yule brought from his claim, near Last Chance, to Michigan Bluff, one thousand seven hundred and forty dollars, as the result of one hundred and thirty-eight days' work, or twelve dollars and sixty cents to the man per day."

Near Coos Bay, Oregon, at a place called Whisky Run, along the sea-shore, in a species of black sand, a deposit of gold was found said to be immensely rich. It is stated that for awhile it was no rare thing for a single miner to take out a thousand dollars' worth of fine gold dust per day.

In 1872,* near Silver City, New Mexico, a boulder was found on the surface, in Lone Mountain District, which weighed two hundred and twenty pounds, and is said to have been worth about two thousand dollars.

A few miles south of Cherokee, Butte County, Cal., were found gravel mines carrying very large nuggets of gold. Pieces were found weighing from one to ten ounces, or from about eighteen to

---

* R. W. Raymond, in United States Commissioner's Reports.

one hundred and eighty dollars, almost every day.

In Yuba County, Cal., the Blue Point Company's ground, an ancient river-bed has paid over one thousand dollars per day's washing, and as much as one hundred and fifteen thousand dollars for less than one hundred days' washings; this, of course, with the aid of a gang of men.

An $8,000 Boulder—A Building Taken Down for the Gold Beneath.

In El Dorado County, Cal., near Pilot Hill, a number of heavy quartz boulders were found, from one of which over eight thousand dollars was extracted.

Bald Mountain, Tuolume County, Cal., has also been noted for its nuggets and coarse gold. Among those found was one of twenty-three pounds, equal to four thousand nine hundred and sixty-eight dollars; one of seventeen pounds, equal to three thousand six hundred and seventy-two dollars, and many pieces of from one to four pounds.

The history of Jamestown, in the same county, is that of many mining camps, of hasty growth and slow but sure decay. The placer mines were formerly of great richness, and for a few years the sales of gold dust averaged about one thousand dollars per day. The bars in the vicinity of Sonoro, Cal., were also formerly very rich. The *Democrat*, of April 15th, 1871, a paper of that place, says: "A brick building is being taken down for the purpose of mining the ground under it. Every day pieces of quartz are found that are very rich in gold. The store was built on ground that had never been mined; it is proving so rich now that a mining hole will soon take the place of the building. Pieces containing from one to three hundred dollars each have been taken out within a week. One week's washing has averaged ten dollars per day to the hand. * * * *

"Several pieces were found ranging from one to three ounces. Twelve wagon-loads, to test the claim before building sluices, paid one hundred and fifty dollars. In the rear of the same building, a few years since, one twenty-five pound chunk was found (five thousand dollars), and several of nearly that weight. A dog digging for a gopher scratched out a piece of gold quartz, which Mr. Condit, the owner of the lot, sold for seventy dollars. Small pieces of quartz are now daily found containing from one to ten dollars in free gold."

Alder Gulch, Montana.

Probably the richest strike, or discovery, ever made in the world, if the after production be considered, was the discovery of Alder Gulch, in Montana.

In 1863, a party of miners on their way to Bannock City, stopped for dinner at the side of a small stream near where Virginia City now stands, and while most of the party were preparing their meal, William Fairweather went to the gulch and panned out a little gravel. The first panful produced thirty cents, and the subsequent ones, about two dollars. As soon as these facts became known there was a general stampede to Alder Gulch from all parts of the territory, and from the whole country at large. At first the production was from one hundred dollars to two hundred dollars per day to the man, so rich were the surface gravels. During the first five years, the gulch yielded *forty millions*, or an average of eight millions per year, and up to the present time, seventy millions have been produced.

Some of the greatest finds the world has ever known have been struck in this territory of Montana. We quote from an address by Mr. R. B. Harrison, who has charge of the U. S. Assay Office at Helena, Montana, delivered before the Bullion Club, of New York city, March 23d, 1880.

Confederate Gulch—One Thousand Dollars to the Pan.

"Confederate Gulch, in Meagher County (Montana), is thirty-five miles from Helena and six miles from the Missouri River, and was worked in 1864 and '65. During the following year, as high as one hundred and eighty dollars to the pan of dirt was taken out near the mouth of Montana Gulch.

"Montana Bar was far richer than the gulch, and

lay at such an elevation above it that it could easily be worked and washed.

"The flumes, on cleaning up, were found to be burdened with gold by the hundred-weight, which was separated at small expense. In the summer of 1866, a drain ditch for Confederate Gulch was projected, and pressed forward to completion in 1868. In the spring of that year, bed-rock was reached, and the enormous yield of one hundred and eighty dollars to the pan in Montana Gulch was forgotten in astonishment at the wonderful yield of *over one thousand dollars to a pan of gravel taken from the bed-rock*.

"Those who have not seen a gold pan will better appreciate this yield, when told that this indispensable article of outfit to the prospector and miner is a foot and a half in diameter, and can hold when heaped, no more than twenty-five to fifty pounds of dirt.

"The average run of a day's sluicing on several of the claims in Confederate Gulch was from three thousand to eight thousand dollars.

"Confederate is not so long or so wide as Alder Gulch, but for its size was far the richer of the two.

"For the same amount of surface, it has produced by long odds a larger amount of gold than any other spot in the world. From it in August, 1866, a four-horse private wagon was loaded with two and a half tons of solid gold, or one and a half millions of dollars, taken from Montana Bar in about ninety days by *three claim owners*. This bar is only half a mile long by two hundred or three hundred feet wide. Every one hundred feet of it produced over one hundred thousand dollars in gold. Some claims, two hundred feet in length across the bar, paid as high as one hundred and eighty thousand dollars to the claim. Where else have we a record of such a marvelous yield of gold? * * * * * * *

Large Gold Nuggets.

"The gold as found in placer mining in Montana varies in size from microscopic powder to a mass weighing one hundred and seventy-eight ounces troy. The latter was the largest nugget ever found in the territory, and came from a tributary of Snow Shoe Gulch in 1865; it was worth three thousand two hundred dollars.

"Another large nugget was found in Nelson Gulch worth two thousand and seventy-three dollars, in July of the same year, and one in Rolker Gulch in 1867 worth one thousand eight hundred. The one now exhibited to the club was found in Deadwood Gulch (Montana) last April; it weighs forty-seven and seventy one-hundredths ounces and is worth nine hundred and forty-five dollars and eighty cents. Large numbers of smaller ones, running in value from nine hundred dollars down, have been found in different gulches. Those from one-fourth ounce to ten ounces are very common even now."

Such statements of wonderful and almost miraculous yields from placer mines, would seem incredible, did they come from a less reliable source; but Mr. Harrison, of the U. S. Assay Office, is unquestioned authority and has had facilities for learning the facts, such as few others have possessed.

Diamonds in California.

Mr. W. A. Goodyear, one of the assistants of the State Geological Survey of California, thus writes to the Placerville *Democrat*, concerning the existence of diamonds in the gravel beds of El Dorado County.

"One other point may be noted as being of some little interest to the miners, as a matter of curiosity, if nothing more, although it is no new thing, it is the occasional finding of diamonds in the auriferous gravel. From all that I have been able to learn it appears that not less than ten or twelve diamonds have probably been found, within four or five miles of this town, and I have no doubt that many more have been picked up and looked at, and thrown away, the finders not knowing what they were.

" During my stay in El Dorado County, I have seen and recognized two of these diamonds, both of which were in the hands of people who did not know what they were, but who had simply saved them as little curiosities on account of their appearance and peculiar shape.

"For the benefit of those who are not familiar with the stone, it may be stated here that this *peculiar shape* of the diamond is one of the easiest and most characteristic features by which it may be recognized.

"The most common shape of the diamond in this country is that of a solid or crystal, having *twenty-four triangular faces*. And another remarkable and easily distinguished peculiarity is that these faces instead of being perfectly flat, as is generally the case with the faces of quartz and other crystals, are very often *curved* or *convex*

Diamonds—How Recognized.

in shape, the centre of each face being a *little higher* than the surface toward the edges.

"The diamond, moreover, is extremely hard, and scratches quartz with the greatest ease.

"If, therefore, any one finds a little white, or yellowish white crystal, with twenty-four of these curved triangular faces; and if, on trying it carefully with a crystal of pure quartz, he finds that it easily scratches the quartz without showing the least abrasion itself, he may be tolerably sure that he has found a diamond."

We have given above a few of the "strikes," or finds, which entice the miners to "rush" to new mining fields, to endure privations, suffer poverty, hunger, and even face death, for the sake of taking his chances in the great lottery which nature has provided for him.

A few, generally but a *very few*, in proportion to the vast numbers who go, reap any lasting benefits. Yet to every new district the old miners are deluded into the same thought again—that this time he will surely win; that this is his "*last chance;*" and hence it is, we think, that so many claims are named "*Last Chance.*"—Hundreds of them of that name.

So with each and every miner that goes, "now," he says to himself, "I shall find my Bonanza." It is a trait of character peculiar to the American people, and especially the American miner, that no matter *how many* reverses he may have met with, or how many sad experiences he may have undergone, he has still the veritable "*pluck*" to try once more. So he "pulls up stakes," and leaves good claims in the hope of finding better ones, and the "rush" is fairly inaugurated.

CHAPTER III.

QUARTZ MINING—VEINS OF ORE—THE MEXICAN ARRASTRA—THE GEOLOGY OF MINING—GRANITE: OF WHAT COMPOSED: DESCRIPTION—METAMORPHIC ROCK—ORIGIN OF MINERALS—WHERE TO LOOK FOR VEINS OF ORE—"FOOLS' GOLD"—IRON PYRITES—QUARTZ VEINS: THEIR PRINCIPAL FEATURES—THE GREAT "MOTHER LODE" OF CALIFORNIA—THE COMSTOCK LODE: ITS SIZE AND PRODUCTION—THE SUTRO TUNNEL: AN ADDRESS BY ADOLPH SUTRO, ITS PROJECTOR—FORMATION OF THE FISSURE: HOW IT WAS FILLED—THE THEORY OF VOLATILIZATION—GREAT DEPTH OF THE COMSTOCK LODE—BONANZAS—CARBONATE ORES OF COLORADO—THE FIRST QUARTZ MILL—THE STAMP MILL.

Quartz.

QUARTZ, as before stated, is the name applied to gold, silver and other ores, in their original state, as placed in the rocks, and is of variable grades of richness, varying from mere nothing in value to thousands of dollars per ton of ore.

The grades most common, and those most frequently mined, yield from ten to one hundred dollars per ton. Nearly all mines have streaks of ore of greater richness than the whole vein will average by mill process, and very few mines are discovered which produce ore for any great length of time that will *average* over one hundred dollars per ton, and most all mines have more or less low-grade ores, which hardly pay for extracting them.

Sometimes gold exists in quartz in almost a

pure state, and is easily extracted; the rock appearing sometimes soft or decayed, and inclined to crumble when exposed to the atmosphere. Thousands of dollars have frequently been extracted from such soft ores, simply by crushing the rock by hand in a common iron mortar, reducing it to a powder, and afterward washing it in a common prospecting pan.

At another place the ore will appear to be alloyed, or mixed, perhaps, with all sorts of base metals, and will be very hard to reduce by mill process, compelling the miner to resort to the "roasting process," and requiring expensive furnaces and machinery, and requiring large capital and great outlay of money to extract the metal from the ore. In some instances the expense has been so great in proportion to the value of the ore, that work had to be discontinued upon the mine, because it was unprofitable, and the miner has had to await the invention of cheaper and better methods of reduction, or abandon his mine as worthless. Such ores are termed *refractory ores*, and are very common in the United States. Frequently these ores are a mixture of silver and lead, carrying some gold. Sometimes copper and tellurium are mixed with the quartz which carries the gold. These metals are found very frequently intermixed in the quartz in various proportions.

These rebellious ores, as they have been termed, were for many years of little value, until the invention of improved methods for their reduction. Finally, after much experiment and oft-repeated failures, the scientific miner gained the victory over them, and at the present time few ores are found which cannot be successfully treated, provided they are of sufficiently high grade to pay the expense.

The Mexican Arrastra.

It is not our purpose in this book to describe all the many processes by which gold and silver are extracted from their ores, therefore only two or three of the oldest and most common methods are noticed. Probably the earliest contrivance for producing gold and silver from the rock, or quartz ore, when found in veins, was the Mexican arrastra. It probably was invented by the Spaniards, or Mexicans, at a very early day, and was brought to California by Mexicans, and is still used to some extent. In Mexico it appears to be still a favorite machine with miners of limited means for the extraction of the precious metals from their ores.

The contrivance, when seen at a distance, somewhat resembles the old-fashioned bark-mill, used by small tanneries, and run by means of a horse walking round in a circle, hitched to the arm of an upright shaft, which revolves slowly around. The foot of this shaft runs in a box or place prepared for it, on a timber imbedded into the ground, and the upper end fastened in a like manner to a frame-work overhead.

Thus far as described, the bark-mill and arrastra are alike. The arm of the arrastra, to which the horse is hitched, is, however, longer than that to the bark-mill, allowing the horse to walk farther from the shaft in a larger circle.

An excavation of perhaps six or seven feet in diameter is made in a circle about ten or twelve inches deep, in the centre of which the foot of the shaft is fixed in place. This excavation is paved in the bottom and on the circular sides with hard, smooth stones—the sides with stones setting up edgewise, or, as frequently made, the sides may be of plank or boards. A solid pavement of stones constitute the bottom; and next, two or more large stones, with one flat side to each, are selected, and fastenings made in them for ropes or chains, by which they are securely fastened to the arms of the shaft, one on each side, opposite each other, with their flat sides resting on the pavement below. These are intended to drag round on top of the paved bottom. The ore is then broken up with a sledge about as fine as the size of an egg, and scattered around on the pavement. A small stream of water is introduced, which can be shut off at pleasure. The horse is started on his rounds, and the work of crushing the quartz is begun.

The ore is thus ground to a fine paste. Quicksilver is introduced occasionally to amalgamate the metal, and when a sufficient quantity has been pulverized, the pavement is taken up, the amalgam collected and washed, and separated from the dirt. The amalgam is collected into a stout canvas bag, and the water is thoroughly squeezed out of it, when it is ready for retorting, a process which has already been described.

A good arrastra will crush from one to three tons of ore in twenty-four hours, and makes from six to ten revolutions per minute. They are frequently propelled by water-power, and attached to a water-wheel and driven somewhat faster.

Such was one of the primitive methods of separating gold and silver from quartz, and strange to say, on ores which were not refractory, it has been but little improved upon, in regard to the quantity of metal saved from the ore, in proportion to the whole amount it contains. In all gold-saving apparatus there is a loss by the fine particles floating off with the water, or failing to be caught by the quicksilver. The arrastra saves perhaps nearly as large a proportion of the metals as the machines of later inventions. Grinding mills and the stamp-mill in due time succeeded the arrastra, as will be fully described later in the work.

Quartz Veins.

With few exceptions, quartz veins are found within granite, slate or porphyry walls, perhaps most frequently within the schists and slates of the primary rocks which lie next to the granite, termed also metamorphic rocks, and occasionally within the rocks termed by geologists the transition group, and the primitive limestone of the primary rocks. It is very seldom, if ever, found in the secondary strata, or those rocks which lie next above the primary.

By primary rocks geologists mean the lowest and oldest formation of the earth, the granite in its varieties being the lowest of this class, and is supposed to extend down to the centre of the earth.

We introduce here a list of the classification of the rocks in the earth's crust, or strata, as given by some geologists, which will aid the general reader in fixing their position in his mind. Beginning at the bottom of the list, the first rock is granite, supposed to be lowest in the earth's strata, and reaching to the centre of the earth; above this the secondary strata, tertiary, etc., with their various subdivisions, until the alluvial or vegetable soil is reached at the surface of the earth. (*See page 103.*)

Granite.

Granite is a close, compact rock, composed of fragments of other rock or stony matter. These are so firmly cemented together, that the whole forms but one solid mass without the slightest indication of pores or fissures. Geologists have

SUPERFICIAL.

Vegetable soil.
Peat.
Gravel beds.
Clay beds.

TERTIARY.

Marl beds.
Shelly millstone.
Gypsum.
Coarse limestone.
Plastic clay.

SECONDARY.

(Chalk Group.)

Chalk.
Green sand.
Weald clays.

(Oolitic Group.)

Oolite.
Sandstones.
Lias.

(New Red Sandstone Group.)

Variegated marls.
Muschelkalk.
Variegated sandstones.
Red conglomerate.
Rock salt

(Carboniferous Group.)

Coal.
Sandstone.
Shale.
Mountain limestone.
Old Red Sandstone.

TRANSITION.

(Granwacke Group.)

Granwacke.
Clayey and sandy slates, or
Lowest fossiliferous.

PRIMARY.

(Inferior Stratified Series.)

Clay slate.
Micaceous slate.
Primitive limestone.
Talcose-granite, or slate.
Gneiss.

(Granites.)

Plutonic—Granite in varieties.

been accustomed to describe this as the oldest and lowest of all rocks, but, in fact, it often appears as a volcanic rock which has been thrown up in a state of fusion through superincumbent strata of other kinds penetrating their chinks, and spreading over them on the surface. These are the peculiar circumstances in which it may be said that other rocks sometimes lie beneath granite.

Granite may then be described as generally forming a base or bed for all other rocks, and as rising in some places from its unmeasured depths into chains of lofty mountains, and in other places penetrating in veins through superincumbent rocks, and partially covering them at the top.

Three substances usually enter into the composition of granite, namely: (1.) Quartz, a gray, glassy substance composed of the oxygen of the atmosphere in union with one of the metallic bases—silicium. (2.) Felspar, also a crystalline substance, but usually opaque and colored pink or yellow, composed of sandy and clayey matter, with a small mixture of lime and potash. (3.) Mica, a silvery glittering substance, which divides readily into thin leaves or flakes, and consisting principally of flint and clay, with a little magnesia and oxide of iron.

In some granites instead of mica we find hornblende, a dark crystalline substance, composed of alumnia, silex of flint, and magnesia, with a considerable portion of the black oxide of iron. Such granites are called *syenite*, from having first been found in the island of Syene.

The matter of which granite is composed is often found to be in the form of small crystals seldom or never assuming the shape of round grains. It is found of all shades and colors, from a bright white to a deep black, often in the same block. The crystals are in many instances not more than one-twelfth of an inch in diameter, but they have been found an inch in size, and even larger.

Granite rock is particularly characterized by the absence of all stratification, or any indication of parallel joints; the rock is uniformly compact in all directions.

Granite rock is frequently interspersed with more or less vertical crevices or veins, which are filled with matter foreign to the rock itself, and sometimes are found to be lodes or veins of ores of various minerals. We may expect to find in these veins ores of tin, iron, copper, lead, quartz, gold, silver and a few other metallic ores.

Metamorphic Rocks.

The rocks of this formation are the second in age. To this class belong a great variety of minerals in rocks covering tracts of great extent and great depth. The rock of this formation is characterized by a partial, and frequently by a decided stratification. It does not belong exactly to either the compact or stratified series.

In this rock, which is very extensive in the United States, we find gold in Virginia, North and South Carolina, Georgia, Alabama, New Mexico, California, Utah, Oregon, and in several other States and Territories. We also find silver in this rock in many places. Platinum is also found along with the gold. Lead and iron ore is found almost everywhere in this formation. This formation is probably the most productive in the useful minerals, and wherever a faint indication of something valuable is discovered in this strata, it is generally worth the trouble to follow and dig after it.

The Stratified Rocks.

The stratified rocks, which are the tertiary and *secondary* divisions of the earth's crust, contains few if any of the precious metals, no lead, copper, nor any metal of consequence except iron and manganese. Precious metals have been found extensively in Utah in a sandstone formation. But it undoubtedly belongs to the *transition* rocks or the sandy slates, in close proximity to the metamorphic rocks.

Coal is everywhere found in the *secondary* rocks in the group termed *carboniferous*.

The upper or tertiary formation though very extensive offers but little inducement to search for minerals, and those few which are found, are generally of an inferior quality.

The Origin of Minerals.

Of the origin of minerals, authors disagree, but a theory very commonly accepted is, that their origin in the form of veins of ore may be considered as the result of infiltration from the surface, to which class many of the iron and copper ores belong, or that the deposits have been formed in the bottom of the sea, as those of the coal measures, or that the minerals have been injected from below, raised by the power of internal heat, to which class the gold and silver ores belong.

One class of veins generally consists of *wedges decreasing* with depth, and another class of spheroidal masses, or pockets, and a third class of *wedges increasing* with depth.

These wedges, described as increasing with depth, are termed by miners, in gold and silver regions, as true fissure veins, and a genuine true fissure vein has never been found to become exhausted as depth is attained, though it is sometimes "pinched" or narrowed, with "*horses*" or otherwise to a very small streak of ore, yet farther down it usually widens out again to its usual thickness.

Hence it is, where true fissure veins are discovered, the only question of successful mining is, in regard to the richness of the ore, such veins having never been known to give out entirely.

From the foregoing it will appear that gold and silver are usually found in regions where the granite or primary rocks have been pushed up to the surface, from great depths below by volcanic action, and are the prevailing country rock of the section in which these metals are found, the secondary or tertiary strata not abounding, this having been displaced or swept away by the elements which exposed the granite and older formation.

Where to Look for Veins of Ore.

It will, therefore, appear that it is nearly useless to look for gold or silver ore, or rich auriferous gravel deposits, in sections where there are no primary rocks to be found; nor is it to be supposed that, because these rocks do abound, that gold can be discovered. Indeed, such rocks are abundant in some localities where no valuable metal has been discovered.

It is safe to say that it will hardly pay the prospector to search for auriferous or argentiferous veins in other than the granite or metamorphic rocks and the older limestone and other formations which belong to the primary class.

There may be exceptions to this rule, of course, but only a practical geologist could determine where the exception exists; and it is not our purpose, in this volume, to give more than a brief outline for determining a gold-producing section.

"Fool's Gold."

It is probable that the substance known as iron pyrites, sometimes termed "Fool's Gold," has deceived very many people. Hitchcock's Geology mentions it thus:

"By no mineral substance have men been more deceived than by iron pyrites, which have been very appropriately named "Fool's Gold." When in a pure state, its resemblance to gold is often so great that it is no wonder those unacquainted with minerals should suppose it to be that metal. Yet the merest tyro in mineralogy can readily distinguish the two substances, since native gold is always *malleable*, but pyrites never. This latter

Iron Pyrites.

mineral is also very liable to decomposition, and such changes are thereby wrought in the rocks containing it, as to lead the inexperienced observer to imagine that he has the clew to a rich depository of mineral treasures. And probably nine out of ten of the numerous excavations which have been made in this country (the East) in search of the precious metal, had their origin in pyrites and their termination in disappointment."

The foregoing describes the iron pyrites of the coal formation. But it appears that the iron pyrites of the primary, or metamorphic rocks, do sometimes contain gold, as the following, from "Overman's Practical Mineralogy," will show:

"Iron pyrites are of little value in themselves. But as a matrix of other metals, namely, gold and silver, they deserve more attention than they have hitherto received. All iron pyrites contain gold, and often silver, from which rule *only those of the coal formation are excepted*.

"The gold deposits of the Southern States constitute virtually a belt or accumulation of veins of iron pyrites. The gold had its seat originally in the pyrites, which, when decomposed, liberate the gold, and it appears in a metallic state. The pyrites are the matrix of the gold."

It appears, therefore, that iron pyrites are valueless of themselves, and not worthy of notice, except when found in the older or gold-producing rocks of the earth.

Quartz Veins.

As before stated, quartz veins, as found in the rocks, vary greatly in the richness and quality of the ores they contain, and also in the ease or difficulty with which the metal can be extracted from the ore. The same is true in regard to the size and thickness and general formation of veins of ore. Sometimes a vein is discovered of but a few inches in thickness, and again another will be of many feet. Frequently their length is limited only by the length of the mountain on which they are found, and sometimes they are only a few rods in length. In regard to the depth which they may extend into the earth, there is the same variation. Sometimes the plane of the vein is but a few degrees from level, and frequently it is nearly vertical. Sometimes they are exhausted of ore but a few feet in depth from the surface; but often they extend down seemingly to the centre of the earth, and are never exhausted, their production being only limited by the expense of working mines at great depth, and the cost of machinery for draining the mine or pumping out the water.

Quartz veins are often of great magnitude. Like the great "Mother Lode," of California, they may be scores of miles in length; and like the mammoth gold lodes of the Black Hills—those on "the belt"—they may reach two hundred feet in thickness; or like the famous Comstock Lode, of Nevada, which has been explored to the depth of three thousand feet, they may reach miles into the earth. Two of the processes of working these veins are illustrated by the opposite cuts.

The great Mother Lode, of California, is a vein, or series of veins, which has been traced on a longitudinal line, with occasional interruptions, for a length of about seventy-five miles, from Bear Valley, Mariposa County, to Amador City, Amador County. Throughout the entire distance it has a general north-west and south-east course, and an almost uniform dip to the north-east of about eighty degrees. In its course it crosses mountains, valleys and rivers, but is nearly one straight line the whole distance. Between its southern and northern extremities it is frequently broken and lost (invariably so at the intersection of the principal rivers), making its appearance again at a distance, frequently in the form of a solid wall of quartz, on the summits of the hills on the line of its strike. These croppings are visible for many miles. It varies greatly in thickness. Frequently, from a series of parallel, narrow veins, it becomes concentrated in one strong, permanent, true fissure vein, of from fourteen to eighteen feet wide.

Usually most all true veins increase in richness with depth; and it is the exception, rather than the rule, that the reverse is the case; yet the Comstock Lode, which has been explored from twenty-two hundred to three thousand feet in depth, and from which about four hundred millions in bullion have been extracted, was the richest near the surface. The first forty tons of ore taken from the Ophir Mine, on the lode, was "packed" on mules, and sent across the Sierra Nevadas to San Francisco, and yielded one hundred and sixty thousand dollars, or an average of four thousand dollars per ton. Yet no body of ore, of consequence, has since been found approaching this value per ton. It may readily be imagined that the discovery of four-thousand-dollar ore created an intense excitement in California.

In some of the mines, the ore now averages from forty to one hundred dollars per ton. The great lode varies in width from fifty to one hundred and fifty feet, and its length is about four miles.

HAND-DRILLING IN THE MINE.

Many theories exist as to the formation of such immense fissures, and as to the causes which filled them with quartz and ore, and there is a great diversity of opinion among scientific men in regard to it.

The theory probably most commonly accepted by practical miners is, that of filling from below, or the pushing up of vein matter from the bowels of the earth, in a liquid state, filling the fissures or cracks in the earth's crust, thus forming these great bodies of ore.

We give below an extract from an address by Mr. Adolph Sutro, the projector and chief engineer

The Sutro Tunnel.

of the celebrated Sutro Tunnel, which pierces the Comstock Lode at a depth of seventeen hundred feet from the surface, and is over four miles in length, being one of the most remarkable and skillful engineering feats ever accomplished in the United States, and which is of incalculable value to the Comstock Mines as a drainage level and means of ventilation.

Mr. Sutro may therefore be considered high authority on this subject, and his views are entitled to great consideration. The address was delivered before the New York Bullion Club, November 6th, 1879.

"The Comstock Lode appears on the surface of a range of hills called the Washoe Mountains, lying east of the Sierra Nevadas and running parallel therewith. The lode occurs mainly at the contact of two kinds of rock, and is therefore in fact, to a large extent, a contact vein, though in other parts, as at the north and south ends, it is surrounded by the same kinds of country rock. The central portion of this mountain range is formed by Mount Davidson, a mountain rising to the height of about seven thousand eight hundred feet, and which consists of syenite; this is probably the oldest formation of that neighborhood.

"Immediately east, and in fact west of Mount Davidson, we find greenstone or porphyry, of which great varieties exist, and which for convenience are called by the family name of prophylite. Still further east we find the trachytic mountain range.

"There have been various theories advanced as to the origin of that lode, but there can be hardly a doubt that it is a true fissure vein; all evidence tends to show that such is the fact.

Formation of the Fissure—Comstock Lode.

"According to Baron Von Richtholfen (who is probably one of the ablest geologists now living, and who has made a careful examination of the Comstock section of country, spending nearly two years there), the syenite is the oldest formation, the prophylite or greenstone coming next in order, while the trachyte is the outburst which appeared at the latest geological period. If we examine the locality, we find, as already indicated, that the Comstock Lode occurs mainly between the syenite and porphyry. The probability is that when the trachyte made its appearance, the upheaval was so great that it uplifted a large portion of the greenstone.

"The effect of this upheaval was, that a fissure was formed at the plane of least resistance, that is, at the point of contact between the two rocks (the syenite and porphyry), large masses of country rock from the hanging wall falling into the fissure, forming what we now call 'horses,' were the cause of keeping the fissure open.

"Had it not been for the fact of these masses falling into the fissure, it would in all probability have closed up again. But in this manner there was left an open channel down to an indefinite depth, which gradually became filled, probably by means of thermal agencies, or possibly by volatilization, according to the different theories which scientific men accept.

Filling of the Fissure.

"These masses or horses must have necessarily fallen into the fissure from above; and as a proof, we have the fact that in the Comstock Lode, every 'horse' consists of greenstone, that being the upper rock, the syenite being at the bottom, none of it could have fallen into the lode. The open spaces thus left in the fissure were gradually filled and the horses became surrounded by quartz and minerals, mainly silver ores carrying more or less gold, which are sometimes accompanied by the base metals.

"I listened with great attention to the lecture of Professor Newburry, delivered last week, in which he expressed the opinion, that the particular fissure which he was describing had been filled in with ore by the process of deposits from thermal waters.

"It seems to me hardly probable that the Comstock Lode was entirely filled in that way. It is probable that different processes were at work at different periods; and it is very likely that a portion of the vein matter which now fills that lode, entered it by the process of volatilization.

Volatilization.

"It seems difficult to imagine silver or gold in a gaseous form; but if you consider for a moment it does not appear strange. We know that all the substances of the entire globe exist in one of three forms; solid, liquid or gaseous; while some substances are familiar to us in all three forms. Take water for instance, we know it as a solid when it is ice, we know it as a liquid ordinarily, and we know it as a gas in the form of vapor. We know all the metals in two of these forms; as solids and liquids when molten. We know some of the metals in all three of the forms. In fact, in our laboratories, we can convert many solids into

The Theory of Volatilization.

liquids by melting, and even into gases by volatilization.

"Now, if we imagine the great laboratory of nature down in the bowels of the earth, where all the agencies probably exist which are necessary for reducing these various minerals to a gaseous state, the filling of fissure veins with metals does not appear so difficult of explanation.

"We must try to realize that in the fact that in that laboratory of nature, there may exist a pressure of millions of millions of pounds to the square inch, and that the steam which is there generated may be heated to a white heat; that is, hot enough to melt iron or any other substance. If we can imagine such a heat as that, we can readily perceive how any substance might be volatilized; and if to these two forces certain chemical agents are added, the transformation will seem still more probable. I doubt that a vein of the size of the Comstock could ever have entirely been filled by deposits from water.

Downward Continuance of the Comstock.

"These theories may be correct or not, but we do absolutely know that we have here a vein which lies between Mount Davidson, the syenitic mountain, and the prophylite adjoining it, extending for a distance of four miles, and reaching downward as far as the miners have gone, and in all probability further than mechanical means will ever permit man to go. There are obstacles in the way which will prevent exploration to an in-

Great Depth of the Comstock Lode.

definite depth. As far as the lode itself is concerned, we find it retains its general characteristics at various depths, that it varies in width from fifty to one hundred and fifty feet, that it consists of solid quartz interspersed with particles of ore; but that in many portions it is not sufficiently rich in ore to pay largely for extracting.

"It seems that the ore of the Comstock Lode often occurs in the form of pockets, or channels, or chimneys, or, as we call them when we find a great ore body, 'Bonanzas.' It is strange that in the vein itself a Bonanza hardly ever occurs.

Bonanzas.

"The lode descends on an incline eastwardly, following the dip of Mount Davidson. In places the pitch is greater than at others, but the average is about forty-five degrees. The ore bodies seem to occur outside and to the east of the vein; they are generally of lenticular form. It frequently happens that in sinking a shaft, or in running a drift,

DEEP MINE WORK.

Surface mining, or washing for precious metals, is the simplest and least expensive form of seeking these valuable commodities. But such are the values of some veins of gold and silver bearing quartz that mines have been sunk over twenty-five hundred feet in depth, involving immense expense and Herculean labors in their working. Virginia City, Nevada, is the great centre of such work.

no ore at all is found; a drift may run right over or under it, while the very next drift may show an ore body of great width.

"This accounts for the great fluctuations which have taken place in the prices of the stocks of mining companies on the Comstock Lode.

"People who are not familiar with the situation do not understand the reason for such fluctuations. But what I have stated will explain one of the causes.

"These ore bodies are not confined to any particular spot. The country to the east of the Comstock Lode may contain ore bodies to an almost indefinite extent. If we imagine, which I firmly believe, that the Comstock Lode continues downward for miles, then it is possible that these ore bodies may make their appearance at comparatively lesser depths, several thousand feet to the eastward of the present workings. The disposition of these ore bodies is not governed by any rule. It seems to be entirely arbitrary. We do not know where they are until we stumble upon them. The only way to look for them is to run drifts all through the country, and then to crosscut from these every one hundred or two hundred feet.

"Some men say that the Comstock Lode is working out. This is nonsense. Several deposits have been found which were of such immense value as to astonish everybody. But these 'Bonanzas' were limited in number, probably not over a dozen altogether; and they were always found in the manner I have described."

From the foregoing it will appear that the value of any vein, when first discovered, is very uncertain. It may be a veritable Bonanza, and it may not be worth anything. The ore may improve in richness with depth, and it frequently does improve as sinking on the vein progresses, or it may be the richest at the surface, or become exhausted entirely.

Even the great lodes like the Comstock, which have yielded their hundreds of millions, are very fickle and uncertain in showing up their treasures.

Carbonate Ores of Colorado.

Frequently they have long periods of unproductiveness, before they stumble upon the rich pockets or shutes of ore which they may contain. Therefore, it is very uncertain how much wealth the discoverer of a lode may have found, and it may take years to fully develop the mine and prove its real value.

Hence it is that quartz mining is usually carried on by large capitalists, by means of great corporations, as the risk and expense is very great for private enterprise. Yet there are many valuable mines successfully operated by private companies or by individuals.

There is another class of ore deposits to which

OVERHAND STOPING.

UNDERHAND STOPING.

we should briefly refer under this head, though a more extended description of them will be found hereafter, in the Colorado portion of this book We refer to the *carbonate ores of Colorado.*

These deposits, when discovered, were almost new to the mining fraternity, and have astonished the world in their magnitude and in their peculiar formation and richness in production. They can hardly be called *veins*, but are vast deposits, lying in nearly a horizontal position, like a coal measure. They have been called contact veins, because lying between the contact of limestone covered by porphyry; but they were so different from the contact deposits of previous history in mining, that miners were some time in learning their true nature and extent. They are deposits of silver ore, mixed with carbonate of lead and iron, lying on a foot-wall or bed of limestone, and covered by porphyry, varying in thickness from a trace to thirty or forty feet of ore—even seventy feet is claimed to have been found of solid ore. They lie in a horizontal position, in places but a few degrees from level, following the waves and depressions of the limestone foot-wall, and covering many square miles in extent, not unlike a coal basin. The ore bodies vary as greatly in richness as in the extent and thickness of the vein matter, and mill, by smelting in furnaces, from ten to one thousand dollars per ton. Even greater richness has been found, but the average is generally from fifty to one hundred dollars per ton. We will refer to this subject again later in the work.

The First Quartz-mill

was built in California, at Grass Valley, Nevada County, in 1851, where quartz mining has been successfully carried on ever since. It was not unlike the stamp-mills of to-day, but, of course, imperfect, and has been greatly improved.

The stamp-mill is the process used on all ores of gold and silver which are free milling ores; that is, ores not requiring to be smelted in a furnace, or roasted, to extract the minerals they contain. Some ores are of such a refractory nature that it is impossible to extract their contents by this process. Such ores usually contain a mixture of sulphur, lead or copper, or all, in addition to the precious metals, in such quantities as to prevent their being amalgamated with quicksilver, in the ordinary stamp-mill, and require smelting.

The Stamp-mill.

All stamp-mills are built upon the principle of crushing the ores to a fine paste or pulp, and thus loosening the metals from their matrix of quartz; they are amalgamated with quicksilver in sluices or batteries with water, after the manner described heretofore.

Stamp-mills are built with from five to one hundred and twenty stamps each, according to the

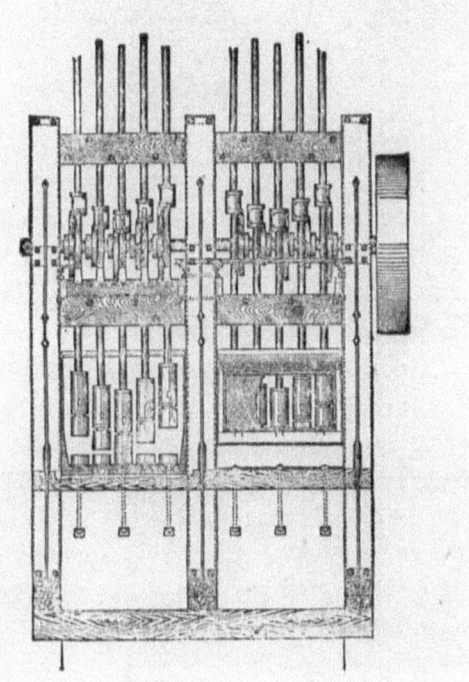

SPERRY'S WROUGHT IRON FRAME STAMP MILLS.

Until within the past two or three years, the cost of wrought iron excluded its use in very many important places. The price at present enables us to make the frame for a Stamp mill of wrought iron, and successfully compete with the wooden frame in price, and make the battery very much better than by the use of wood, especially as in most instances the timbers of which a battery frame is to be made, are cut from the forest and used almost immediately, without the least seasoning. When such timbers are used, although put together by the most skilled workmen, they shrink and open at the joints. Soon the whole structure becomes shaky; the nuts upon the bolts jar loose, requiring constant attention. These defects are all overcome by the use of wrought iron in the construction of the battery frame. This material being slightly elastic, successfully prevents the transmission of the jar through the frame, and by the use of washers of the same material the nuts will not jar loose from the bolts.

capacity required. These stamps may be termed shafts, or bars of iron, standing perpendicular, weighing from six hundred to nine hundred pounds each, their feet or lower ends shod with steel, which drop into iron mortars or boxes, also plated with steel. These stamps are raised by means of "cams" or eccentrics, placed on a horizontal shaft of iron about five inches in diameter, which lifts them up from eight to eleven inches, when they instantly drop into the shoe or mortar with great force upon the ore placed therein.

The ore, after being crushed in a rock-breaker to the size of an egg, or smaller, is introduced into the shoes, into which a stream of water is running, and the stamps drop upon it at the rate of sixty strokes per minute, for each stamp, crushing it to a fine paste. In this state the ore passes off through a screen and on to the sluices or blankets, over copper plates or into pans, according to the plan of the mill and according to the kind of amalgamating apparatus used, which collect the gold and silver. The mills are usually "*cleaned up*" every week, and the amalgam retorted, as in hydraulic mining.

A good stamp-mill will crush nearly two tons of ore to each stamp in twenty-four hours. They are driven by both steam and water-power.

As the location of veins and placer claims, and the laws and regulations governing their location, are intimately associated with the subject of mining, and as a knowledge of the subject is indispensable to those seeking Government title to mineral lands, in the next chapter will be found the most important laws on that subject.

Before leaving the subject of stamp-mills and pulverizers, the following explanation of the Phelps "Little Giant" Stone and Ore-Crusher (of which pictures are given on the opposite page) will be of interest and profit:

EXPLANATION OF OPPOSITE CUTS.

A. Side Plates.
BB. False Crushing Plates.
C. Wrought-Iron Side Bars.
DD. Set Screws and Plates.
F. Reciprocating Jaw.
G. Rock Shaft.
H. Toggle.
I. Liners, or backing, by means of which the Reciprocating Jaw is adjusted to crush fine or coarse.
J. Main Shaft.
K. Eccentric Pitman.
L. Eccentric.
M. Fly Wheel.
N. Spring.
O. Pedestal.
P. Bearing of Reciprocating Jaw. (Dust proof.)

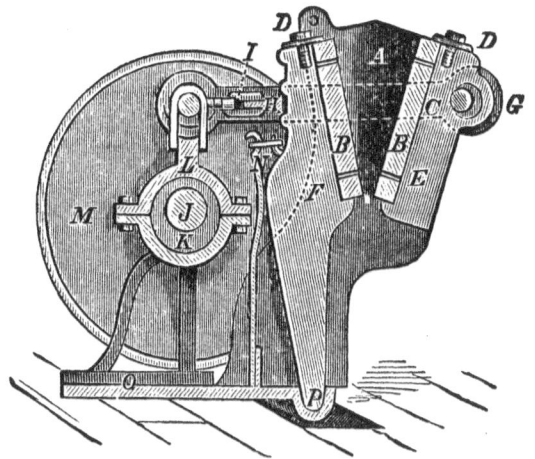

SECTIONAL VIEW.

THE "LITTLE GIANT" STONE AND ORE CRUSHER.

CHAPTER IV.

MINING LAWS—HOW TO LOCATE MINERAL CLAIMS.

The Size of a Legal Claim, etc.

IN the early days of gold mining in California, and in other States and Territories, on lands belonging to the United States, the mining laws were very imperfect, and disputes about the titles of claims were of constant occurrence.

It appears that upon the discovery of every new gold field, that the miners met and organized what was termed a "Mining District," fixing certain boundaries for it, and adopting a code of laws and regulations governing the size of claims within the boundaries named. Therefore, each new district, although generally following after previous codes, was liable to have a set of laws quite different from other localities, and entirely dependent on the fancy of the miners who organized it. Accordingly, the size of a claim that miners could hold varied greatly in different localities—such being the crude state of the laws in those days—and very naturally disputes about title were very frequent.

In the early days there were no courts convenient to settle such disputes, and a class of miners called "Jumpers," taking advantage of this fact, became very numerous. If a miner found a rich vein or a lucky strike, he was not at all sure of holding his claim unmolested from this lawless class. If he left his claim for a short time, he might return to find another in possession, who would dispute his title if ever so good, and as there was no authority at hand to expel such intruders, except the law of *physical force*, men frequently took the law into their own hands, and generally the strongest party held the claim.

From this cause there was much wrangling, and frequently bloodshed and murder were of common occurrence. Jumping claims, if not so common now as in those days, is still in vogue, and such occurrences are common in all new mining regions, though to a less extent than formerly, and probably always will be. There can hardly be any law or authority strong enough to entirely protect the weak against the strong, when great selfishness and lust for gold are the leading characteristics of the community.

According to the revised statutes of the United States, no vein or lode claim made subsequent to May 10th, 1872, can exceed in size a parallelogram fifteen hundred feet long by six hundred feet wide. But whether surface-ground of that width can be taken, depends upon the local regulations of State or Territorial laws in force in the several mining districts. But no local regulations of State or Territory can limit a vein or lode claim to less than fifteen hundred feet in length along the course thereof, whether the location be made by one or more persons, and surface rights cannot be limited to less than fifty feet in width, unless adjoining claims previously located on each side render such limitation necessary. The end lines of all claims must be parallel to each other.

The Cost of a Government Patent.

The owner can follow his vein anywhere in all its dips, spurs and angles, providing he does not go outside of his end lines.

In Colorado, the Legislature passed an act, in 1874, that all locations made thereafter in the Territory should carry with them surface-ground of one hundred and fifty feet on each side of the centre of the vein, or in other words, could not exceed in size a parallelogram three hundred feet by fifteen hundred feet, except in the four counties of Boulder, Gilpin, Clear Creek and Park, where seventy-five feet on either side of the vein should be the rule, making a legal claim in these four counties of half the width above given. The law, however, does not interfere with lodes discovered prior to their adoption.

Previous to the enactment of this law, claims had been of various sizes at different periods. At one time, the discoverer of a lode could hold but two hundred feet on the vein, and subsequent locators but one hundred feet on the same vein. Afterward it was changed to fourteen hundred feet for the discoverer, and for a short time it was enacted that the discoverer could hold three thousand feet on the lode.

After a Government patent has been issued for a lode claim, the owner can hold it against all comers or claimants, whether he works it steadily or allows it to remain idle.

The cost of procuring a Government patent is from one hundred and twenty-five to one hundred and fifty dollars, when the location embraces one thousand five hundred feet by one hundred and fifty feet, and from twenty-five to thirty dollars more than this, when fifteen hundred feet by three hundred feet are included. The former class embrace about five six-hundreths acres, and the latter about ten thirty-three-hundredths acres, which, at five dollars per acre, the price fixed by the Government, is not far from twenty-six and fifty-two dollars for the land. The other expenses are twenty-five dollars for the surveyor-general; from thirty to forty-five dollars for the deputy surveyor; twelve dollars and fifty cents for certified copies and abstracts; ten dollars for filing; eighteen dollars for advertising in the newspapers, and fifteen dollars for notary fees, all of which have been made sufficiently large in this estimate.

The laws provide that five hundred dollars worth of work shall be done upon a claim before application for a patent can be made.

The laws of Dakota Territory, governing the

size of claims in the Black Hills country, provide that a vein or lode claim shall be in width one hundred and fifty feet on each side of the centre of the vein, or that its surface-lines cannot exceed a parallelogram of three hundred by fifteen hundred feet.

In regard to placer claims, their size is in such a measure dependent upon the local regulations that we omit to give more than the United States mining law in regard to it.

During the session of 1871–72, Congress revised the statutes in regard to the location of mineral claims upon the public domain. Those specially interested in these laws will do well to apply to the Department of the Interior for particulars.

The method is, however, stated below in brief.

Any person, association or corporation having claimed and located a piece of land for mining purposes, may file in the proper land office an application for a patent, under oath, showing their compliance with the mining laws, together with a plat and field-notes of the claim, as made under the direction of the United States surveyor-general, showing accurately the boundaries of the claim, which shall be distinctly marked by monuments on the ground. They shall post a copy of such plat, together with a notice of their application for a patent, in a conspicuous place on the land embraced in such plat previous to the filing of their application, and shall file an affidavit of at least two persons that such notice had been duly posted, and shall file a copy of the notice in the land office, and shall thereupon be entitled to a patent for the land, in the manner following: The register of the land office, upon the filing of such application, plat, field-notes, notices and affidavits, shall publish a notice that such application has been made, for the period of sixty days, in a newspaper to be by him designated as published nearest to such claim; and he shall also post such notice in his office for the same period. The claimant at the time of filing this application, or at any time thereafter, within the sixty days of publication, shall file with the register a certificate of the United States surveyor-general that five hundred dollars' worth of labor has been expended or improvements made upon the claim; that the plat is correct, with such further description by such reference to natural objects or permanent monuments as shall identify the claim, and furnish an accurate description, to be incorporated in the patent. At the expiration of the sixty days of publication the claimant shall file his affidavit showing that the plat and notice have been posted in a conspicuous place on the claim during such period of publication. If no adverse claim shall have been filed with the register at the proper land office at the expiration of the sixty days of publication, it shall be assumed that the applicant is entitled to a patent, upon the payment of five dollars per acre.

CHAPTER V.

THE BLACK HILLS: EARLY HISTORY AND DISCOVERY—INDIAN TRADITIONS—ORIGIN OF THE DISCOVERY OF GOLD—THE FIRST PARTY TO WINTER THERE—EVIDENCES OF FORMER OCCUPATION—THE FIRST CITY—THE FIRST DISCOVERY OF QUARTZ—COST OF SOME OF THE MINES—ANNUAL PRODUCTION—MINING REGULATIONS, ETC.

The Black Hills—Early History and Discovery.

ABOUT two hundred and forty miles north of the Union Pacific Railway, in the midst of an alkaline desert, and embraced between two branches of the Cheyenne River—a tributary of the Missouri—rises a magnificent mass of mountains, covering an extent of territory about as large as the State of Vermont. To these mountains the Sioux Indians gave the name of "Pah Sappa," which, interpreted, is Black Hill. These are the Black Hills proper. Another range of mountains, about two hundred miles west and south-west, are called by this name, and are lain down on some maps as such, but they properly belong to the Laramie Mountain range, and are a continuation of these mountains, and should not be called Black Hills.

The Black Hills rise up abruptly and alone in the midst of a plain more than two hundred miles from any other range of mountains, their highest peaks reaching about seven thousand feet above sea level. They might truly be called an oasis in a wide and dreary desert, its approaches on every side being through long stretches of treeless plains, whose waters are so strongly impregnated with alkalies as to be unfit for the use of man. But when once the hills are reached, all this is changed. Dense forests abound; springs and streams of pure water are abundant, and along the creek bottom-lands grows luxuriant grass. The soil of the valleys is rich and fertile, and it will become a fine grazing country.

The Black Hills' Indian Natives.

It was the opinion of the exploring expedition of 1875 that the Hills had never been the permanent home of the Indians. Had the country been used as such a residence within a period of thirty years, it is claimed some marks of its occupation would have been visible. But the party learned that, although small bands of Indians go a little way into the Hills to cut lodge polls, all signs indicated that they were the merest sojourners of the most temporary sort. These views were corroborated by the Indians themselves. A few of them, through curiosity at the presence of the soldiers, came into the Hills in 1875, and one old chief conversed with the interpreter. He said he was fifty years old, and had always lived in that section of country, in the vicinity of the Hills, but that he had never ventured into them before; that the

BAD LANDS MOUNTAIN.

"Bad lands," is a title given very generally by the Indians to any sandy or clayed tract which is specially barren. The sand hills of Dakota, in the Black Hills country, are known by this name. A stretch in Southern Utah and Northern Arizona bears the same designation.

squaws came sometimes to cut lodge polls, but their stay was very short; that the Hills were "bad medicine," and the abode of spirits; that it thundered and lightened, and rained very hard, sometimes; that the Indian does not like rain; that the lightning tears the trees to pieces and sets fire to the woods; and, moreover, that the Indian had never lived there.

It is said these statements were borne out by every Indian communicated with, and they are, undoubtedly, in a great measure true.

Origin of the Discovery of Gold.

The following are said to have been the causes which led to the discoveries in this region:

Some Indians came into a frontier trading post, bringing small grains and nuggets of gold. Plied with presents and "fire-water," they said it came from the Black Hills.

Other accounts of the discoveries in the Black Hills say that as far back as 1849, Indians exhibited specimens of gold to trappers and hunters, which they claimed to have found in the Hills.

It appears that as early as 1855, General Harney visited the Hills on a sort of an exploring expedition, and the highest peak has been named in his honor (Harney's Peak). General Warren visited them in 1856-57; Dr. Hayden in 1858-59, and General Sully in 1864. Father Desmet, a Catholic missionary, after whom the Desmet Gold Mine was named, also visited the Black Hills at a very early day.

From the reports and observations of these many explorers it seems that the Hills were believed to contain great mineral wealth, although but little had been accomplished in the way of discovery. It is asserted that in 1852 a party of men, en route for California, influenced by reports of gold in the Black Hills, made their way thither, and found rich diggings; but that they were all massacred by Indians except one, who made his escape, but died soon afterward of disease. It seems some evidences tending to confirm this report have been found in the remains of old rotten and decayed sluice-boxes, and marks of mining works found there at the time of the great rush in 1876–77.

These rumors spread, and becoming greatly exaggerated, all classes were excited by them. To solve the problem, the Government sent out Custer's expedition, of 1874. After his return, the question still being undecided, the expedition of 1875, under the charge of Professor Jenney, was ordered by the Interior Department, at Washington, of which some account has been given.

This expedition found gold in many places, but no placer fields of very great richness; in fact, none that would pay the ordinary pan or cradle miner. The position of the present rich mines in the vicinity of Deadwood was still undiscovered,

BETTER RESULTS. 145

Professor Jenney did not visit Deadwood and Whitewood Gulches, owing to the dense forests and timber which presented obstacles to his entering with his train.

In 1876, these gulches were prospected by miners and found to be rich in placer deposits. During that summer the population was swelled by an immigration of nearly seven thousand, who settled in and around Deadwood Gulch, and the yield of gold during the year, from the placer mines alone, was one and a half millions of dollars. Quartz mining was not yet at this time inaugurated.

Just previous, however, to the advent of Professor Jenney, and during the winter of 1874–75, a party of miners, consisting of twenty-two men and one woman, made their way into the Hills by following the trail by which General Custer came out in '74, and camped on French Creek, at a place

The First Party to Winter in the Hills.

since known as Camp Harney. Here they erected a stockade of upright logs, set two feet in the ground and rising twelve or fourteen feet high. This was made eighty feet square, with flanking projections at the corners, of very straight logs, ten inches in diameter, set as close together as possible and battened on the inside. There was but one gate, and that was on a side difficult to approach, on account of the creek, and was barred by a bullet-proof door. Inside this inclosure were five log cabins, and between the cabins and wall a space sufficient for their wagons and animals. These miners had also laid out a town, and the foundations of several houses were begun.

This party, after all this labor had been expended by them, were compelled to leave, and were brought out of the Hills by troops about March 1st, 1875. They had been there but four months, yet in that short time, in addition to building the stockade, had riddled the valley with prospect holes and had worked on quartz leads in places far into the Hills. These were the first miners who ever wintered in the Black Hills, of which there is any reliable record. The fortress made by them has been called Tallant's Stockade, from Colonel D. G. Tallant, who is thought to have been the leader of the party, and his wife, Mrs. Tallant, being the first woman to enter the Black Hills.

Evidences of Former Occupation.

This party claimed to have found evidences of former occupation by white men. In the bed of the creek was found the rotten remains of cottonwood trees, of which the heart had been hollowed out to form sluice-boxes. They found the remains of two shovels, pieces of tin and fragments of iron not far from the decayed sluices. They also claimed to have found the remains of an old

MINERS AROUND THEIR CAMP-FIRE.

cabin, formed by poles leaning up against an overhanging rock, and not far away a wooden cross, standing upright on a rock, tied together with rawhide, and near its base, cut into the moss-covered stone, were two rude letters, "J. M.," and the date, "1846." This was supposed by Tallant to mark the grave of some unfortunate pioneer. How reliable these stories may be, the author has no means of ascertaining, but gives them for what they are worth.

By the 20th of July, 1875, during the stay of the Government expedition, at least six hundred miners were engaged in mining and prospecting. As this country was then an Indian reservation, the Government desired that the miners should be taken out, and accordingly General Crook came in person to Custer City, called a meeting of miners, to take place August 10th, 1875, and requested them to voluntarily leave the Hills. He conversed freely with them and urged them to depart, making known that his orders were imperative to remove them from the Indian reservation.

The First City Organized.

Before going out, however, the miners proceeded to organize a "city." A beautiful site was chosen in the valley of French Creek, a commission appointed, streets laid out and named, lots surveyed and numbered, and all else having been done, a difficulty arose about choosing a name.

The miners were about equally divided between Northern men and those from the South. The Northerners wished to call the town "Custer City," and the Southern men wished it to be "Stonewall." Debate ran high, words nearly led to blows, weapons and revolvers were freely brandished; finally a vote was taken, and when the judges of election arose, after counting the ballots, they pronounced the name "Custer City."

The ownership of town lots was decided by each man drawing from a box a folded ticket, on which was written the number of the lot of which he was to become the owner.

This town organization took place August 10th, 1875. Two other towns were laid out during the same summer—one on Spring Creek and the other on Rapid. But neither arrived at the dignity of a house. Custer City, however, boasted of two large log structures, one intended for a courthouse and the other a hotel. These constituted the towns of the Black Hills, when General Crook brought the miners out in 1875.

Early in 1876 Custer became a "*booming*" camp. Its streets became crowded with business houses, tents, log huts, dance-houses and hotels; but the placer mines not proving very rich, the population

First Discovery of Quartz Lodes.

scattered over the country, and upon the discovery of the richer deposits of Deadwood Gulch, later in the season, Deadwood City sprang into life, and became the objective point of the great immigration that was pouring into the Black Hills, and Custer City became almost deserted.

During the summer of 1876, the population of the Hills was swelled to a number not less than seven thousand, most of whom settled in and around Deadwood City.

Other towns were formed, and "Gayville," a mile and a half above Deadwood, and "Crook City" (after General Crook), eight miles below on Whitewood Creek, were laid out simultaneously. "Hill City," on Spring Creek, and "Rapid City," on Rapid Creek, were formed a little earlier than Deadwood. Central City, two miles up Deadwood Creek, and Lead City, four miles up Whitewood Creek, which streams unite at the city of Deadwood, were built about a year later, and became towns of considerable importance, because lying in the vicinity of some of the great gold mines of the Hills, at which there was subsequently large quartz-mills erected.

It was an auspicious day for the Black Hills when a miner happened to uncover a ledge of singular-looking iron-stained quartz down on the bed-rock of his gulch claim, near the present site of Deadwood. The specimen fairly sparkled with golden grains. He ground it to powder on the flat surface of a large stone, washed it in a prospecting pan, and found it to contain gold at the rate of nearly one dollar for every pound of rock. Druggists' mortars were then brought into requisition, and miners began to pound out from these claims considerable amounts of gold. Quartz locations then became in order, and developments proved the existence of immense deposits of gold-bearing quartz in veins of such great size as were heretofore unknown, and quartz mining and prospecting was begun, and has been very actively prosecuted since. This was the spring of 1877.

California capitalists were not long in learning of the great size of these deposits, together with their richness and free-milling qualities, and the ease and moderate expense with which the ores could be manipulated, and in the winter of 1877–78, they made heavy investments of capital in quartz lodes in the vicinity, and made preparations for the erection of large quartz-mills for the reduction of the ores. These capitalists paid four hundred thousand dollars for the "Father Desmet" mine; eighty thousand dollars for the "Golden Terra;" for the "Homestake," No. 1, seventy thousand; "Homestake," No. 2, fifty thousand, and for the "Old Abe," two hundred and fifty thousand.

These mines have since erected mammoth mills. The Homestake built a mill of eighty stamps, following it by the erection of another of one hundred and twenty stamps, being, it is said, the largest quartz stamp-mill in the world. The other large companies have erected mills nearly or quite as large, so that the present annual production of gold from quartz alone in the Black Hills has reached the immense sum of about five million dollars.

The gold yield of 1877 is estimated at two million five hundred thousand dollars; that of 1878, three million five hundred thousand dollars, and that of 1879, at nearly six million dollars. Undoubtedly, however, these figures are large, and probably overestimate the amounts produced rather than undervalue them. It is probable, however, that 1880 will fully reach a production of six millions or over for the Black Hills mines.

Black Hills Mining Regulations.

Below will be found a few points of the mining laws of Dakota Territory, in connection with both general and local statutes governing the location of mineral claims:

Only citizens, and those who have declared their intention to become such, can legally locate mines.

All land is mineral that is more valuable for mining than farming purposes.

A vein or lode extends one hundred and fifty feet on each side of its centre, and the end lines must be parallel with each other.

Where two or more veins intersect or cross, the prior location will take the ore within the intersection.

The discoverer must record within twenty days from the date of the discovery, and his location certificate must contain (1) the name of the vein; (2) the name of the locator; (3) the date of location; (4) the number of feet in length claimed on each side of the discovery shaft; (5) the number of feet in width claimed on each side; (6) the general course of the vein, as nearly as may be.

Any cut deep enough to disclose the vein, or a ten-foot adit or trench along the vein from the point of discovery, would be a legal discovery shaft.

All mining claims are subject to the right of way, for mining purposes, of any ditch or flume, tramway or packtrail in use, or that may be laid out across such location; but such right of way shall not be exercised against any location made prior to the claim of such right of way without the consent of the claim owners, except by condemnation, as in the case of land taken for public highways; and such ditch or flume shall be so constructed as not to injure vested rights.

No location certificate can embrace more than one claim, no difference how many locators there may be. The register of deeds is entitled to one dollar and fifty cents for recording and furnishing a certified copy—one dollar for the former and fifty cents for the latter.

CHAPTER VI.

WESTWARD HO!—NOTES BY THE WAY, ETC.

WESTWARD ho, over the snow-capped Rockies! was our exclamation, as we left our pleasant home in Pennsylvania, to seek our fortune and cast our lot for a time in the Far West. Yet it was not such a light and joyous exclamation after all; it meant a world of things to us; it meant leaving a happy home and loved ones, dear parents and friends, wife and babes, to endure the hardships and suffer the trials of a long and tedious journey in a new and unknown country, to brave the dangers of wild camp-life in the wilderness, and perhaps added to these was the terror of molestation from Indians and road agents up among the mountains. Thus, with all our enthusiasm down deep in our hearts, covered by this outward calm, we carried a load of sadness.

But we had little time for meditation; our train was at hand, and with a hearty "good-bye" and a "last look" from the platform, we were soon, ere we were aware of it, passing out of sight of the grand old hills of Pennsylvania.

To us living in the Middle States it seemed as if all railroads led to Chicago, and there seems to be hardly a single route that does not lead to that city, and to reach it people will only need to take the road nearest at hand, or, in other words, "go it blind," and in twenty-four hours they will be rolling into the metropolis of the lake country in the West. But from Chicago to the Far West Eastern people must choose a route. We chose the Chicago, Burlington and Quincy Railroad, and we think it offers to travelers going to all parts of the Far West a greater variety of routes and better connections perhaps than any other. The influx of people this season into Kansas, Colorado and the mining regions was simply enormous, and this road to accommodate this travel had already been running three through express trains daily (with through cars attached, without change) between Chicago and Council Bluffs, Omaha, St. Joseph, Atchison, Topeka and Kansas City. They had also found it necessary to put on a new fast train with through cars between Chicago and Kansas City, reaching its destination in twenty hours, making in all four through express trains daily.

At Chicago we took one of Pullman's sixteen-wheel sleeping cars, and lay down at nine P. M., and went to sleep; when we awoke the porter was shaking us, and exclaiming: "The next station is Burlington—twenty minutes for breakfast!" The sun was shining brightly, and it was not far from six o'clock. We had enjoyed a splendid night's rest, and were two hundred and seven miles from Chicago.

Before we had found sufficient time to make all these observations, we were crossing the bridge that here spans the "Father of Waters." Surging, swift and muddy, it rushes beneath, indeed a mighty river. There is a draw in this bridge across the Mississippi, to allow steamers to pass.

Soon we were within the limits of Iowa, and the newsboys, before the cars had fairly stopped at the station, were crying: "Morning papers," "Daily *Hawkeye*," and we remembered that the *Hawkeye* man, who here resided, had become famous by means of this self-same sheet. Burlington is rather pleasantly situated upon a bluff which here rises in places to considerable height, overlooking the Mississippi, and has a population of about thirty thousand.

But these iron horses seem never to tire or grow weary. Ours gave us little time to study the scenery by the way, and we were soon rushing along again over a rolling prairie, with a rich, black soil, and here and there patches of timber, and frequently traversed by streams and rivers. West from Burlington our road lay for awhile through the apparently richest farming country we ever saw. Farmers were plowing and marking for corn, and the sulky plow, as well as the ones of older pattern, were everywhere in use.

All along we saw evidences of the greatest thrift and enterprise.

In our car were a number of stock men, or drovers, mostly very plainly dressed and apparently hard-working farmers from Iowa, yet each wealthy in the number of their cattle and swine, as compared to many of our Eastern stock-growers. We found conversation with these men to be very interesting, and learned not a little that was new to us in this line. These men had been to the Chicago stock-yards, with each from two to five car-loads of steers apiece, or from thirty to one hundred head of cattle each. These steers average them about forty dollars, net; therefore, these men were taking back with them from one thousand two hundred to four thousand dollars to the man. These trips to Chicago they were constantly making; it was their business, followed every week, at which they were making good profits. They had, indeed, "struck a lead," and will probably "find it rich." They told us the receipts in Chicago the previous day had been unusually large, some seven thousand head having been brought into the stock-yards. The average receipts are said to be from three to five thousand. The cattle raised in this section are made to weigh much heavier than average Eastern cattle. They told us it was very common for two-year olds to weigh from eleven to fourteen hundred pounds. They had just seen a cow in the stock-yards that weighed two thousand one hundred pounds. As we heard it from the mouths of several, it was probably true.

We learned, also, something about that terrible pestilence among swine—the hog-cholera. One told of a contract made by a neighbor to deliver sixty hogs at a certain price to a dealer on the following Monday—this being on Friday. When the time arrived only one of the hogs was alive, all having died of cholera save this one, and five of the fifty-nine having died the very day the contract was made. Another stock man said that he had "learned how to manage the business;" that when he first began he "was green," and lost his hogs by letting them "die on his hands;" but now, as soon as he found they "were drooping," he shipped them at once and "sold them." He furthermore said that unknowing ones could not tell the difference, but, with his experience, he could tell by appearances when the disease was coming on. Goodness gracious! we thought, is this the *honor* of Western pork-raisers, and does such pork ever get into the barrels we buy in the East branded from Chicago packing-houses, and unconscious of its being any but the sweetest and purest of meat. But he had something more to tell us. When the cholera rages badly, there are parties who make a business of buying the carcasses of the dead hogs, gathering them up in large numbers at their factories, where the lard is extracted for grease; the hogs costing mere nothing, but the grease selling for enough to allow them to realize large profits, it being used by soap manufacturers, and also for machine oil. In fact, when purified and whitened, it is difficult to distinguish it from pure lard; and they told us, moreover, that in refining the grease operators were in the habit of tasting (?) it at a certain stage of the process, to see if it were tasteless and free from odor, or properly "purified" (pardon us for threatening to forbear eating pork or lard hereafter.)

But we now had to bid most of our drover friends adieu, as, in the meantime, we had traversed nearly the width of the State, and were nearing Council Bluffs. As the sun was sinking behind the western horizon we crossed the muddy waters of the Missouri at Council Bluffs, and entered Omaha at about eight o'clock P. M., having come five hundred and two miles from Chicago in good time, and being not a tedious or tiresome journey for so long a distance. Looking at our watch, which was Philadelphia time, we found it to be half-past nine o'clock, or an hour and a half too fast. In going westward with the sun, we had lived an hour and a half for which the calendar will never give us credit! An hour and a half *gained,* in which we have lived and moved, and yet of which we have no record, save the hands of the faithful time-piece in our vest.

Tired and dusty, we stopped for the night.

Omaha is nine hundred and sixty-six feet above the sea-level. It is a dusty, dirty, yet busy city. A large portion is well built with brick and stone, and contains many elegant and costly edifices. It has, in addition to its retail trade, wholesale houses representing all branches of business, and according to her board of trade report, does a wholesale business of fifty millions annually. It has single firms who carry stocks of goods to the amount of two hundred and fifty thousand dollars, and which claim to do a business of one and a half millions annually. The city also has large manufactories of various kinds. The Omaha Smelting Works have one hundred and seventy-five thousand dollars invested in grounds, buildings and machinery; they employ from one hundred and fifty to two hundred and fifty men, and their monthly pay-roll amounts to ten thousand dollars. During the year 1879 this firm produced in fine silver and gold from the ores the value of four million dollars besides producing nineteen million pounds of lead.

We had little time, however, to view the many points of interest in the city. The day following our arrival we boarded a Union Pacific train of fifteen or more passenger coaches, which were crowded and packed full to overflowing with human beings of every grade, and station, and nationality, all bound toward the setting sun. Already we were getting within the exciting influences of the great gold and silver regions of the Rocky Mountains. Here we encountered a party bound for the Black Hills—young men, enthusiastic over the prospects of digging gold in the golden Hills. A little later we overheard a gentleman conversing with a passenger about the silver mines of Colorado. The party said he preferred the San Juan country to Leadville, and to use his words: "There you have a vein to follow, while at Leadville you have nothing but a flat deposit to be worked out."

In all these far western journeyings a most fraternal sociability prevails. Old men and young, established capitalists and roving adventurers, chat together in the utmost freedom. One needs to watch sharply lest he be imposed upon.

Our train, after passing through a rolling country near Omaha, soon brought us into the great Platte Valley, which presents to the eye one of the smoothest, most level plains which the mind could imagine. Here is a valley from ten to twenty miles wide, of great length and extent, and frequently, in some directions, as far as the eye can reach there is not a tree or an object to obstruct the sight. Fences are almost unknown, though occasionally a barbed-wire fence may be seen in the vicinity of towns. In the older portions of the country a few trees have been planted, but very few are to be seen anywhere. Timber being so scarce and lumber undoubtedly so high, the houses are small, and barns there are none—at least, they are few and far between. Small sheds, covered with straw, take their places, and sod houses and "dug-outs" are the substitutes for houses. Very frequently corn-cribs are seen filled with corn, but without roofs or cover of any sort. It was very dry here, and we were told it had been six months or more since rain had fallen.

Farther west we reached the great cattle ranges, where we saw vast herds of cattle and sheep grazing in the distance. Water appeared to be very scarce, and of very poor quality; cattle were seen drinking from holes in which the water was of inky blackness. The soil is a rich black loam, like the soil of Illinois and Iowa. Windmills are always in sight, used in pumping water for stock; at one point we saw a large one, evidently used in grinding grain. Hence it seems that this ancient invention has become new again, and is of very practical utility in more senses than one.

"Prairie Schooners"—The Great Cattle Ranges.

Soon the white "Prairie Schooners" began to come into view, all sailing westward. We passed many of these canvas-covered wagon trains in companies of three or four, all pointed toward the setting sun. Inwardly we exclaimed, "All the world is going West!" All day long over the same level plain, and all night the same, only lighted up by frequent prairie fires, of which we were never out of sight, sometimes lighting up the heavens with a lurid glare, presenting a grand and almost frightful spectacle. Daylight found us at Ogalalla, three hundred and forty-two miles from Omaha. Here it was that the Union Pacific train was robbed a couple of years ago, by a daring and desperate band of highwaymen, of considerable treasure and express matter. A great change had come over the face of the country. Instead of the plowed fields and green meadows farther east, there were very few evidences of civilization. The grass was parched and dry, covered by shifting sands and alkalies. Still, the herds of cattle in the distance had grown larger. The plain was everywhere marked with cow-paths, crooked and winding, made by the cattle passing to and fro to water; occasionally the bleached skeletons of cattle were seen, white and decayed. This is the section where the cattle are turned loose in winter to shift for themselves and seek their own subsistence; some die, but it is said the majority come out in the spring all right. Look where you might, on either side there were always cattle in sight in this section. Frequently we saw antelopes galloping away from the track across the plain.

About eight o'clock, A. M., we reached Sidney, Nebraska, distant from Omaha four hundred and fourteen miles, and its elevation being four thousand seventy-three feet—we having raised in elevation since leaving Omaha over three thousand feet, so gradually as to be hardly perceptible. Sidney has a population of about one thousand.

From here the Sidney and Black Hills stage line runs daily stages to Deadwood, two hundred and sixty-seven miles distant, making the trip in about fifty-six hours. Fare to Deadwood, thirty dollars; by through ticket much less. We saw this stage loading with baggage at the stage depot, preparing for a start to the Hills; although apparently already overloaded with boxes, trunks, valises and packages of every kind, they were still tying on mail sacks and luggage in every conceivable place where it was possible to fasten a bundle. The place has become of considerable importance as a freighting and outfitting point for the Black Hills gold-fields, and has large forwarding houses and freighting companies to supply the Black Hills trade.

The stage-ride from the Union Pacific Railway, at Sidney, to Deadwood, can hardly be called a pleasure trip, and over a dreary and barren plain through furious clouds of dust, filling eyes, and ears, and nose, it soon becomes very monotonous, and a ride of fifty consecutive hours by day and night becomes tiresome in the extreme. Yet, with all, it is a new and novel experience, full of adventures to those who have never taken such a journey, and whirling around a sharp precipice on a narrow track after a four or six-horse team, will perhaps cause no little concern as to safety, and, perhaps, much admiration for the skill which can guide the leaders so dexterously. If, in addition to the other trials of the journey, the route be infested with "road agents," or stage robbers (as was this route at one time, an account of which will be given), the tax on the nerves is a severe one.

But when the vicinity of the Black Hills is reached, and the barren plains are succeeded by noble streams and rich soil, and luxuriant grasses take the place of alkalies and sage brush, it is a most refreshing change.

At Rapid City, which is about fifty miles from Deadwood, the stage reaches a rich valley, with good farms and ranches, and fields of waving grain. Numerous houses dot the plain, and an appearance of civilization, which is a glad reminder of Eastern homes and farm-houses, is shown upon every side.

About thirty-two miles farther on the stage halts at Fort Meade, the Government military post, situated in a lovely valley near Bear Butte Creek, and eighteen miles beyond, after traversing the roughest and most picturesque part of our route, the metropolis of the Black Hills, Deadwood, is reached.

CHAPTER VII.

BLACK HILLS—DIVISION INTO COUNTIES—STREAMS—RIVERS—CREEKS—
SCENERY — MOUNTAINS — PEAKS — FRENCH CREEK — SPRING CREEK—
RAPID CREEK—BOX ELDER CREEK—BEAR BUTTE CREEK—WHITEWOOD
CREEK—SPEARFISH RIVER—GAME, ETC

Black Hills Scenery, Streams, etc.

THE Black Hills proper embraces all that country lying between the "North Cheyenne," or "Bellefourche" River, on the north, the Cheyenne River on the south, and Wyoming Territory on the west. These two rivers running eastward unite and form the Big Cheyenne, making, with the Wyoming Territory line, an irregular triangle containing between five and six thousand square miles of territory, out of which three counties of Dakota have been formed. The western or end lines of these counties lie next to Wyoming, and their longest or side lines run due east and west, Custer County on the south, of which Custer City will be the county seat, Pennington County lying north of it, of which Rapid City is the capital, and Lawrence County the farthest north, lying next to, and bounded by the Bellefourche River on the north, of which Deadwood is the county seat. These are the local subdivisions of the Black Hills country.

Extending through these counties from north to south a distance of about seventy miles is a series of high and rugged

Mountains,

which attain their highest elevation on the north-west or near the Wyoming Territory line. These mountains rise to an elevation of from five to seven thousand feet above sea level. But the country is so broken, and they are so nearly of equal height, that excepting one or two peaks, no one mountain seems to rise much above the others.

Harney's Peak, in Custer County, attains an elevation of seven thousand four hundred feet. Terry's Peak and Crow, in Lawrence County, reach a height of seven thousand two hundred feet and six thousand two hundred feet respectively, and are noted peaks. Bald Mountain, a little south-west of Deadwood and Bear Butte Peak, about fourteen miles north-east of the same, are prominent land marks.

This series of mountains, which occupy an extent of country about ninety miles long by sixty wide running north-west and south-east through the counties named, have a central or granitic portion within the limestone belt which sorrounds them at their outer " mesa " of about seventy miles long by thirty to forty wide, irregular in shape but within which all the gold and silver deposits are found.

As before stated, these mountains attain their

BASALTIC COLUMNS IN THE GREAT NORTH-WEST.

highest general elevation on their western rim, and hence, most of the streams of the hills flow in an eastern direction, and empty into the Cheyenne River.

Some of these streams, in cutting their way out through the eastern *mesa* or foot-hills, have worn deep cañons and wonderful gorges through the mountains on their way to the plains below.

After passing out on the plain and reaching the dry and thirsty soil, most of these beautiful streams become swallowed up by the thirsty sands and are lost. They sink beneath the surface and with one exception only (Rapid Creek) they fail to reach the river during the summer season.

French Creek.

French Creek, in Custer County, of three hundred miner's inches, rises on the western rim and flows eastward for fifteen miles through a beautiful, elevated park country, then plunges into a deep and rugged cañon, from which it does not emerge for twenty miles, until it reaches the plain and soon sinks out of sight. Custer City is situated on this stream at a point before it leaves the beautiful park country we have described.

French Creek is well supplied with timber, and the soil in the park country named is well adapted to agriculture. The stream becomes very small during the summer, some seasons very little water running in it during the dry months of the year.

Considerable gold is found along the stream, north of French Creek, and in Pennington County is *Spring Creek*, of seven hundred miner's inches. It runs eastward by a tortuous course through deep and narrow valleys and rugged cañons, and finally cuts its way through the great eastern mesa and is lost on the plain. With a fine, deep bed, it carries no water out of the hills. Gold is found in considerable quantities along this stream and its tributaries, and it has ample water for sluicing. It offers but little inducement to the agriculturist, the valleys being small and the country very moun-

Spring and Rapid Creeks.

tainous. Rockerville is situated a few miles from its banks and the town of Sheridan on this stream. Pennington County has also two other large streams, Rapid Creek and the Boxelder. The former, of two thousand miner's inches, is one of the largest streams in the Black Hills, and the only one that cuts through the eastern rim which escapes the thirsty secondary, and pours eastward across the plain to the Cheyenne River. Many small streams unite to form it, and among these, Little Rapid and Castle Creeks are the most prominent. The gold diggings on Castle Creek have been among the richest of the Black Hills. Rapid Creek also passes through a deep cañon on its way out to the plain. Rapid City is situated on this stream at a point whence leaving the cañon the stream enters a lovely valley or plain which offers one of the choicest spots in this country for the settler. There are many other towns on this stream and its branches, among them, Rochford, on Little Rapid Creek, and Castleton, on Castle Creek. From source to mouth it is a fine stream and probably the most valuable for all purposes of any in the Hills.

The Boxelder Creek,

of three hundred miner's inches, lies a few miles north of Rapid Creek and runs eastward parallel to it. It has some silver and copper deposits within the territory it drains, but no gold discoveries nor towns of importance along its banks.

Bear Butte Creek.

Bear Butte Creek rises in the northern rim of the hills in Lawrence County, a few miles south of Deadwood, and runs north-east close to the mountain from which it takes its name and empties into the Bellefourche.

The town of Galena and the rich silver deposits there, are on the headwaters of this stream, also the Government Military Post of Fort Meade, and Sturgis City further down the stream. It carries but a few miner's inches of water.

Whitewood Creek,

in Lawrence County, rises within the northern mesa in the vicinity of Bald Mountain and Terry's

Peak, and runs northward into the Bellefourche. It has several small branches, the principal one being Deadwood Creek, from which that city took its name, and which stream unites with the Whitewood, within the limits of the city. This section was formerly heavily timbered with pine and other woods along this stream, and was not mentioned in the report of the expedition of 1875. The deposits of gold in this gulch was the immediate cause of the great rush into this section in 1876.

We have only space to mention one more stream of the many remaining ones of the Black Hills.

Spearfish Creek Scenery.

The Spearfish, also in Lawrence County, is a stream perhaps even larger than Rapid Creek. It rises in the northern mesa in the vicinity of Bald Mountain, runs north for many miles through the grandest, deepest cañon of the Hills, a chasm in many places not less than two thousand feet deep, from which it emerges out on a valley or "bottom" of the richest soil more than a mile wide, the finest agricultural valley of this county. This valley is about seven miles long to where the Spearfish empties into the Redwater River, which, ten miles farther on, empties into the Bellefourche. Spearfish City is on this stream, and is the centre of the most flourishing agricultural portion of the Black Hills—no gold is found along this stream. A project is planned for taking water from the headwaters of this creek and carrying it in a ditch fifteen or twenty miles to the vicinity of Deadwood, for the purpose of working the placer mines by hydraulic process. However, it is not yet completed.

There are a few gold deposits on the headwaters of streams flowing through the northern mesa, but the streams are small ones—Potato Gulch and Bear Gulch are among these, and from these gulches some gold has been extracted.

The scenery along some of these streams and deep mountain gorges is magnificent. Foaming, dashing streams of clear, cool water rushing through cañons so deep and narrow as almost to exclude the sunlight, except at noon, whose sides are so precipitous as to offer no foothold for man or beast, are common scenes in the Black Hills. A specimen of these appears opposite.

Beautiful parks between hills, covered with luxuriant grasses and carpeted with lovely wild flowers are frequent on the headwaters of French and Spring Creeks. To the tourist or pleasure-seeker they offer the grandest of sights and one of the fairest resorts in the world.

Game.

When the Black Hills were discovered they were bountifully supplied with game, and elk, deer, bear, antelope and smaller game were found in abundance. No doubt the genuine sportsman can still find here plenty of opportunities to try his skill, but unquestionably the sturdy miners have long since driven away much of the larger game or killed it for food.

SPECIMENS OF BASALTIC COLUMNS.

CHAPTER VIII.

DEADWOOD CITY—EARLY HISTORY—LAYING OUT THE TOWN—THE FIRST CABIN—A BIG SALE OF GOODS—THE GREAT FIRE—RISING FROM ITS RUINS—REMARKABLE ENERGY IN REBUILDING—DEADWOOD REBUILT—THE FIRST POST-OFFICE—STAMPS SOLD—MONEY-ORDERS ISSUED—COST OF LIVING IN DEADWOOD—DEADWOOD MARKET REPORT.

Deadwood City—Early History.

ON the 25th day of April, 1876, in a wild and picturesque gorge, nearly five thousand feet above sea level, the town of Deadwood was laid out, at the junction of Whitewood and Deadwood Creeks.

Some parties claim that a man named "Fyler" was instrumental in laying out the embryo city. Others say that Craven Lee, Isaac Brown, Noah Seiver, James Hamilton, E. Durham, Charles Barber, J. J. Williams and "Red" Young laid out the town, using a lariat rope instead of a surveyor's chain for a measure. However, the town was laid out, and the first house, a pine-log cabin, was built by Lee and Brown, about the first of May, and by the middle of June the place had grown into a bustling town. At that time the buildings were of the poorest character, nothing, in fact, but tents and log cabins; even the stores and business places were of the same kind. Everything bore the aspect of a mere camp in the woods. Stumps and trees blocked the streets in every direction; a forest had just previously covered the site. Gradually the timber disappeared, and by the first of July three steam saw-mills were turning out about twenty thousand feet of lumber daily, and were not sufficient to supply the demand for building and mining purposes. Every industry seemed to prosper. Merchants who went there in 1876 and early in 1877, taking with them stocks of goods, disposed of them at fabulous prices. One man took into the place in his own wagons goods which he bought in Cheyenne for three thousand dollars, and sold them as fast as he could open them from the boxes, receiving for them about ten thousand dollars.

The village of 1876 grew into a town in 1877, and a rushing city in 1879. Handsome public and private buildings were erected on every hand; five hundred shops and mercantile houses, some of them carrying from twenty-five to fifty thousand dollars' worth of goods, were scarcely sufficient at this time to do the business of the place. There were three daily papers, three banks, hotels and restaurants almost without number, three theatres, four churches, and a fine graded school-house. On September 25th, 1879, Deadwood had put on metropolitan airs, and boasted of over five thousand inhabitants. By the next day the fire-fiend had leveled it to the ground.

The Great Fire.

About half-past one o'clock, on the morning of September 26th, a fire broke out in a bakery on Sherman Street. The building and those adjoining it were of the most inflammable character—wooden buildings, made of yellow pine. The wind blew a gale, and sparks flew in showers, setting fire to everything they touched. The flames roared through the business portion, and on to the private residences along the hillsides, sparing nothing in its course.

The startled citizens had barely time to escape with their lives, many of them saving only the scanty clothing which they caught up as they fled. The blowing up of some houses with giant powder stopped the fire's devastating career at China Town, and the tearing down of other buildings at the intersection of Pine and Sherman Streets checked it there. But it was too late. From the corner of Pine and Sherman Streets to China Town, an area of nearly half a mile long by a quarter wide, every house, whether brick or frame, with the exception of about a dozen small fire-proofs, was gone. In three hours from the time the fire began there was not a dry-goods, grocery, or boot and shoe store; not a hotel, theatre, bank or printing-office left in what had been the main business part of Deadwood. The destruction was complete and the loss total. Nearly three hundred buildings and their contents were swept away, and daylight dawned upon two thousand houseless and homeless people, perching around half naked on the bleak mountain crags, with heaps of ashes and cinders as their only possessions on earth. Thus, in an incredible short space of time, over one and a half million dollars' worth of property was destroyed, on which there was scarcely any insurance.

But, phœnix-like, the ruined city, in a few short months, as if by magic, has arisen from its ashes. It is said that merchants, while their buildings were still in flames on that memorable morning, were galloping out before daylight to the saw-mills to order lumber and materials for building, and that the foundations of new structures were lain while the stones were still hot from the effects of the fire. A month later a Deadwood paper stated: "The whole place is a teeming hive of human bees without a drone. The streets are blockaded with wagons, teams, lumber, brick, mortar and throngs of sturdy laborers. Every man you meet has his coat off and working tools in his hands. The whole air is filled with the clatter of hammers and saws. Men with lanterns swinging on the dirt banks beside them, dig away all night long on excavations for cellars and foundations; and carpenters, with miners' lamps on their hat-fronts, nail on boards and laths and shingles at midnight as busily as noonday. Sundays, as well as week days, the din and rush go on without ceasing. The whole population exchanging, lending and helping each other; law-offices in bar-rooms, courts held in shanties, and newspapers interchanging type, labor and material. Amid all the loss and desolation there has been no wrangling about titles, though every record was destroyed, and possession is almost the sole evidence of proprietorship; no disorder, no whining, no sign of discouragement. And to-day Deadwood, nearly rebuilt, stands forth the cheeriest, pluckiest little city in all Christendom."

Deadwood Rebuilt.

At the present time (six months afterward) but very few traces of the great fire remain. The new town which has arisen is an improvement on the old one, the buildings are larger, finer and better made, and business has in every way resumed its natural channels.

To give an idea of the vast transactions going on in this little mountain city: As early as 1877 one bank did a business of from twenty-five to seventy-five thousand dollars *per day*, in buying gold dust and selling exchange, and one hotel at this time fed as many as one thousand people inside of twenty-four hours; single firms were selling goods to the amount of from seven to ten thousand dollars per month. This business, of course, became divided up later, as the town filled up and competition became greater, but on the whole, the trade of the city largely increased.

Deadwood is now the distributing point for many outlying towns and mining camps, and of the twenty million pounds of freight which arrived in the Black Hills during 1879, probably over two-thirds of it came to Deadwood, and was distributed from there.

The Post-office—Cost of Living.

The first post-office was erected by R. O. Adams. The business grew rapidly, and a larger building soon had to be provided, and about six hundred lock-boxes were put in the new building. At present a daily mail is received from each of the routes *via* Sidney, Cheyenne and Bismarck, and a weekly from Fort Pierre. About six hundred pounds of mail matter is brought in daily by these stages, and about eight thousand letters are received every twenty-four hours. Two hundred dollars' worth of stamps are sold daily; six hundred and fifty dollars' worth of money-orders issued, and about five hundred dollars' worth sent in registered letters daily. The amount of money-orders issued in 1878 was one hundred and thirty-five thousand nine hundred and seven dollars and fifty-seven cents.

The cost of living in Deadwood is from eight to

twelve dollars per week at the principal hotels or boarding-houses. Prices of provisions of all kinds are, of course, much higher than in the East, as the cost of freighting them over such long distances adds from four to six cents per pound to everything brought in. The average price of flour during 1879 was about eleven dollars per barrel; potatoes, one dollar per bushel; eggs, forty cents per dozen; bacon, sixteen to eighteen cents per pound; hams, seventeen cents per pound; pork, per barrel, thirty dollars; corn, about one dollar and ninety cents per bushel, and oats about one dollar and forty cents per bushel, wholesale prices. Owing to the increase in home productions, produce is much cheaper than formerly.

Below will be found a list of Deadwood retail prices, July 6th, 1879:

Deadwood Market Report.

PROVISIONS.

Potatoes............per lb., 1½	Eggs............per doz., 35
Beans, Navy............ " 8	Honey............per lb., 30
Cheese............per lb., 20@25	Lard, in pails............ " 20
Butter, common............15@20	Bacon, common...per lb., 12½@14
" good ranch............25@30	Hams, smoked..... " 17

CASE GOODS.

Peaches............2-lb. can, 25	Pine Apple............2-lb. can, 30
Corn............ " 30	Jellies............ " 50
Strawberries............ " 35	Star Lobsters............1-lb. can, 30
Raspberries............ " 25	Peas............2-lb. can, 25
Tomatoes............ " 25	Salmon............1-lb. can, 30

DRIED FRUIT.

Apples............per lb., 17	Blackberries............per lb., 35
Peaches............ " 20	Cherries............ " 35
Currants............ " 17	Raspberries............ " 50
Prunes............ " 17	Raisins............ " 30

TEAS.

Imperial............per lb.	65	Japan............per lb.,	30@60
Hyson............ "	65	Green............ "	60@$1
Young Hyson............ "	65	Oolong............ "	65

COFFEE.

Rio, good............per lb.,	30	Rio, prime to C............per lb.,	28
" common to fair..per lb.,	25@28	Java, Old Government... "	40

OILS AND PAINTS.

Head Light............per gal.,	$1.00	Linseed Oil............per gal.,	$1.50
Lard Oil............ "	1.30	White Lead............per lb.,	16½

SUGARS.

Cuba............per lb.,	18	New York Coffee, A......per lb.	18
New York Crushed... "	22½	" Extra, C........ "	16

SYRUPS.

New York Syrup.....per gal., $1.25 | New York Drips...5-gal. kegs, $6.50

FISH.

White.............per kit, $2.50 | Mackerel............per kit, $2.75

BREADSTUFFS.

Flour, State............per 100, $7.00 | Flour, Dak............per 100, $8.00

GRAIN.

Corn ....per 100, $5.50 | Oats............per 100, $5.25

SALT.

Fine Dairy..........per bbl., $14.00 | Coarse.... ....per bbl., $12.00

Deadwood, unlike many mining towns, is at present a quiet, orderly city; much of the wickedness and dissipation of the early days has disappeared. It seems to be a law-abiding community. It is true that in the early days murders and bloodshed were frequent. But the stranger who comes there now will hardly see a trace of such character at present.

CHAPTER IX.

BLACK HILLS GOLD MINES—THE FIRST MINES DISCOVERED—THE FIRST QUARTZ-MILL: WHO BUILT IT—THE RENO MINE—RICH ORE—CEMENT MINES—MONTHLY YIELD OF SOME MINES—FOUR THOUSAND SEVEN HUNDRED LOCATIONS IN THE HILLS—VEINS OF ENORMOUS WIDTH— "THE GREAT BELT:" THEORIES ABOUT ITS FORMATION—THE DESMET MINE: ITS PRODUCTION—THE DEADWOOD MINE: ITS PRODUCTION—THE HIGHLAND MINE: COST—THE HOMESTAKE: ITS COST: PRODUCTION—THE GIANT AND OLD ABE: COST—THE RHODERIC DHU MINE—PRODUCTION OF SOME OTHER MINES—DESMET MINE AND MILL—THE HOMESTAKE MINE AND MILL—THE CALEDONIA—THE GOLDEN TERRA.

Black Hills Gold Mines.

QUARTZ mining and prospecting in the Black Hills appears to have begun almost simultaneously with placer mining. During the year 1876 many locations were made. The "Father Desmet" mine was discovered in May, and the "Sir Rhoderic Dhu" on the 23d of October following. Very soon afterward the "Homestake," and most of the other best-known mines.

Subsequently, the great size of the veins becoming known, as development progressed, a general interest in quartz prospecting was the result, and hundreds of locations were soon made. The first quartz-mill dropped its stamps on January 1st, 1877, and was built by Robert Lorton an M. E. Phinney. It was a custom-mill.

Another class of deposits besides true veins began to be developed, known as cement de-

The Reno Mine.

posits, and were found to be exceedingly rich. In February, 1877, the "Reno" mine, of this class, was discovered by L. G. Turner, A. Sawdry and A. L. Louden, near Central City. Within one short year some four thousand dollars or more were produced from the mine, all of which was pounded out of the rock in a common hand-mortar, and washed out in an ordinary prospecting-pan. The ore for a time was valued at fifty dollars per ton, and the owners were offered one hundred thousand dollars in gold for the property, and refused it. A single ounce of ore from this mine yielded three dollars and fifty cents, and one and a half pounds of rock, crushed with a sledge-hammer, yielded the astonishing sum of sixteen dollars. These were selected specimens, and of course not a fair test of the whole deposit, but it all added to the interest and excitement in quartz mining at that time. The Reno produces now about eight thousand dollars per month.

The other cement mines best known are the "Great Eastern," which produces about seven thousand dollars per month (and which claims to have struck a true vein in one of its shafts). The "Aurora Consolidated," which puts out about seven thousand five hundred dollars, and th

"Gustin," producing about one thousand two hundred dollars monthly. The ores from these mines are treated like that taken from true veins, being crushed in stamp-mills and amalgamated.

From the vicinity of Deadwood Gulch miners extended their operations to other localities, and soon everywhere, all over the Black Hills, within a radius of seventy miles, very promising locations were made. Up to May, 1879, over four thousand seven hundred locations of quartz mines had been recorded, and new discoveries have been reported almost daily ever since.

The Great Belt.

The oldest and best-known mines are situated on what is termed "the Great Belt," a mammoth lode of several miles in length, situated between and extending from the vicinity of Central City on the north, south to Lead City. The vein is of enormous width—from forty to one hundred and fifty feet in various places, and continuous in length several miles.

The Country Rock, of which the mountains through which it passes is formed, is a micaceous slate, one of the older or primary rocks, which here has been thrown up on its edge, the lines of its stratification bearing at an angle of fifty or sixty degrees from the horizontal. Between these lines of stratification have been formed great fissures, which have been filled with vein matter in solution, forming the lode described.

It is not our purpose to discuss the theory as to how these veins were formed. Professor Jenney, of whom mention has been made, holds that these are true fissure veins. "Interlaminated fissures," that is, fissures opened between the layers of the rock, and not across the line of stratification, as is frequent in other localities. The auriferous quartz, he says, has been formed by water solutions, which have come up from below; and believing them to be true fissures, he supposes them to extend into the earth to an indefinite distance, and that they will increase in richness with depth.

Other scientific men claim that they are not true fissures, but merely deposits formed in aqueous solution with oxide of iron, and precipitated between the walls of the rock while it was still in a horizontal position before its final upheaval to its present nearly perpendicular position. If the latter theory be correct, the veins will of course be liable to become much sooner exhausted than if the theory of Professor Jenney be right.

But, allowing either theory to be correct, the amount of good ore in sight is immense, and it will take many years to exhaust the apparent supply now opened up.

The great mines of the Black Hills, as known at present, are mostly on this "Great Belt," described above. Other districts may become in time fully as well known, and will undoubtedly prove as rich, but we have only space to describe those best known at present.

"The Belt" extends in a northerly and southerly direction some distance west of Deadwood, and from north to south. It is something over two miles in length. In several places this great lode is proved to be fully one hundred feet wide, and in others nearly twice as wide. Beginning on the north end of the lodes, the first mine fully proved

The Desmet.

to be on "The Belt," is the "Father Desmet." However, the "Sir Rhoderic Dhu," a valuable mine just north, is believed by many to be on the same great vein, and in time may prove to be so in fact.

The Desmet mine is owned by Messrs. Archie Borland, August Hemme, L. R. Graves and A. J. Bowie, Jr., all of California. The ground owned by them comprises three claims, "Father Desmet," "Golden Gate" and "Justice." The claims all being three hundred feet wide by one thousand five hundred feet in length on the lode, but two of the claims (the Desmet and Golden Gate) lying parallel to each other, forming a parallelogram six hundred feet wide by one thousand five hundred feet long. The owners of these three claims only have a total of three thousand feet in length on the Belt. It is probable that separate organizations or stock companies will be formed to work each of these claims.

The Desmet is now producing about sixty-one

thousand dollars per month, and has large mills in course of erection. The ore averages about twelve dollars per ton by mill process, and the claims cost the owners four hundred thousand dollars, which is the highest price paid for any claim or set of claims in the Black Hills.

The Deadwood Mine.

South of the Desmet group is the "Deadwood" mine, which is owned by Messrs. Lloyd Tevis, J. B. Haggin, George Hearst and Gilmer, Salesbury & Co. The claim comprises something over one thousand feet on the lode. The company have an eighty-stamp mill, which produces about forty-five thousand dollars per month. The expenses per month being only about eight thousand dollars, the company is making a net profit of nearly thirty-seven thousand dollars monthly.

South of the Deadwood mine is the "Golden Terra" claim, of five hundred feet on the Belt. It has two stamp-mills, one of twenty and another of eighty stamps. The product of the mine is about sixty thousand dollars per month.

Lying next south is the "Highland" mine (which is sometimes called the "Homestake, No. 2"). It comprises four claims, the shortest one being eight hundred feet long. It is owned by Messrs. Tevis, Haggin & Hearst, and Gilmer, Salesbury & Co., and cost them two hundred and fifty thousand dollars. It is building a one-hundred-and-twenty-stamp mill, which, when it is completed, will produce about seventy thousand dollars per month. The present product is fifteen thousand dollars per month.

Lying south of the Highland, is the Homestake, which owns two claims, the Homestake and Golden Star, each of which is one thousand five hundred feet long by three hundred feet wide. It is an incorporated company, its shares selling on the market at all the exchanges, and cost seventy thousand dollars for the Homestake claim and fifty-five thousand dollars for the Golden Star. It has two mills, one of eighty and one of one hundred and twenty stamps, the latter, at the time of its erection, being the largest gold-mill in the world in the number of its stamps. The product of the mine is about one hundred and fifteen thousand dollars per month.

The Giant and Old Abe, the next mine south, owns a strip of territory equal to one hundred and fifty feet long on the belt, by six hundred feet wide, and cost the corporation who bought it two hundred and fifty-two thousand dollars for all the conflicting interests. The mine is sufficiently developed to be safely called upon the belt, and is erecting a one-hundred-and-twenty-stamp mill. These locations comprise the mines known to be on the same great ledge, although, as stated, others may prove to have found the lode with depth after further exploration has been made.

Among the valuable mines not on the belt, the Rhoderic Dhu produces about twelve thousand dollars per month; the Esmerelda, ten thousand; Durango, ten thousand; High Lode, thirty thousand; Gopher, ten thousand; Caledonia, which is a very valuable property, forty thousand per month. North of the Father Desmet mine, and south of the Homestake, the great mother vein seems to dip into the earth, and is apparently lost, although extensions are claimed to be found, and may become so, in fact, which further developments will have to demonstrate.

The Desmet Mine and Mill.

The Desmet Mine, as originally purchased, comprised the Justice, Belcher and Golden Gate, all lying parallel to the claim whose name now designates all. The "Justice" was the first of these to make the acquaintance of the prospectors' pick in 1876.

The Desmet proper was opened by removing the surface dirt from the ledge, which came to the surface in many places, showing up a "face" of sixty feet of ore when the sale was made. The new owners pushed their development from this breast, already exposed, finding, when a wall was encountered, that the ledge was one hundred and fifty-four feet across. An occasional barren boulder, or "horse," was met with, but from this excavation twenty thousand tons of ore were extracted and milled, which yielded an average of ten dollars per ton. Crossing the ledge, as defined by the wall already exposed, a body, or chimney of rich quartz, about four feet wide, paid eighty dollars per ton, probably the richest ore, where any great quantity exists, yet found in the Hills.

The mountain has been penetrated by a drift for a distance of four hundred feet to a point near its centre, and the same character of ore (not, however, this eighty-dollar ore) is found. Crosscuts at different places along the line of this drift prove the ledge to be at no point narrower than seventy feet, while the greater portion of it is much wider. The mountain side being very steep, drifts four hundred feet long penetrate a long distance below the surface, and as the overlying mass in this instance is known to be chiefly ore, the quantity in sight above this drift will be readily comprehended as immense. It is asserted that the vein of ore showing up in the Desmet, and from this mine through to the Homestake, a distance of one and one-fourth miles, is the largest vein of ore yet found in the world, when its great width is taken into consideration. The course of the ledge is slightly south-west and south-east, with a dip to the east of about thirty-five degrees. At the surface-opening near the "Belcher," the quartz is one hundred and fifty feet in width, with ore beyond. The vertical shaft of the "Justice" is down in ore one hundred and seventy-five feet;

thence by winze of one hundred and forty-seven feet to a tunnel, where the ore from the upper workings is dumped into cars, and transported along the tunnel three hundred and forty feet to an automatic elevator, situated about forty feet from the mouth of the tunnel. The cars are there lowered on the elevator to an inclosed tramway, running direct to the mill. The tramway is two hundred and eighty-one feet long, and crosses from the hillside to the mill with a span of over forty feet in length, and is at an elevation of seventy feet from the surface. At the end of this tramway the cars are emptied into rock breakers, and thence, by means of chutes, runs to the bins, from which it is fed to the stamps. The eighty-stamp mill is perfect in all its arrangements.

The principal work on the mine is now confined to the two-hundred-and-fifty-feet level. Here an immense chamber, over one hundred feet square, has been excavated, the south-west side showing the foot wall of the vein. The breast of ore exposed is over one hundred feet wide, hard and firm in its character, impregnated with white iron, and mills well in gold. The ore has very little waste intermixed with it, and averages from twelve to twenty dollars per ton.

The Homestake

was among the first Black Hills properties bought in the interest of California capitalists, having been

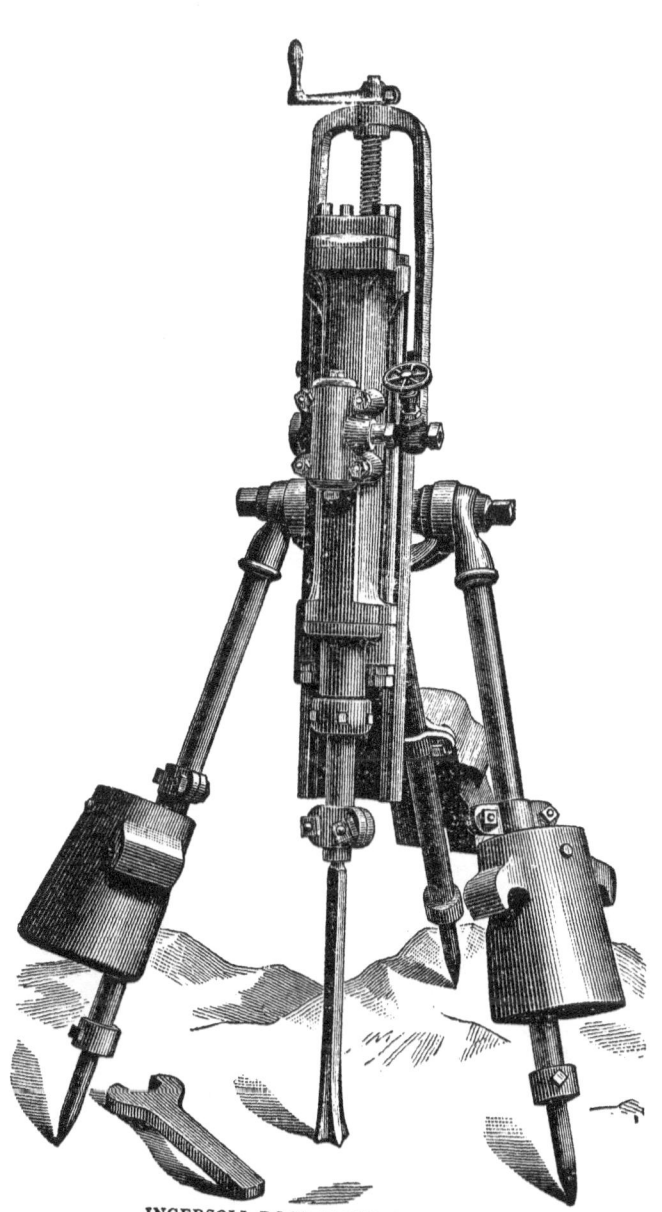

INGERSOLL ROCK DRILL ON TRIPOD.

purchased by George Hurst, in conjunction with J. W. Gashwiler and Henry Janin, early in the month of April, 1878. The mine, which embraces a number of valuable claims, covers an area one thousand five hundred feet long by four hundred and fifty feet wide. The many tunnels and shaft-openings in every part of it give evidence that the Homestake is one great, continuous body of paying ore. The most systematic and thorough developments have been made to test the real value of the property, and the result has been the erection of an eighty-stamp mill, and another of one hundred and twenty stamps, and the preparation to work the mine by all appliances that will procure a steady yield of bullion. The north drift, from the vertical shaft on the one-hundred-feet level, is in over one hundred feet. It shows solid ore of a very rich character. In the Golden Star cut the breast has been widened. At one time a large body of talc slate, near the surface, was exposed to view, which assayed over fifty cents to the pound. How extensive this body may be is not known. At all points of the mine late developments show an increase and richness of the ore.

Over two hundred car loads per day are now run out over the tracks, thirty-four men working in the mine. The shots or blasts are enormous. Holes are opened by the aid of the best steam drills, and charged with enormous amounts of block powder, which tears down over a hundred tons of ore.

Five hundred and twenty-five tons of ore per day is the average amount crushed. But if the cubic measurements made by mining experts are correct, there is enough ore now in sight to run five hundred tons a day for over five years. By means of an elevated tramway, over fifty feet above the ground, and supported by trestle-work, a track of "T" rails is laid into the top of the large one-hundred-and-twenty-stamp mill. An engine traverses this track from the mine to the mill, hauling as many cars of ore as may be desired. The building is one hundred and sixty by ninety feet. The sides thirty-two feet high, and the height from floor to the cone of the roof is seventy-eight feet. The framework throughout the entire structure is of the best selected Black Hills timber. Over one million feet of lumber was used in the construction, and one hundred and sixty-eight thousand shingles.

The celebrated Corliss engine, of three hundred horse-power, shipped from Providence, R. I., is the largest ever brought into the territories. It weighs eighty-nine thousand pounds, and has two fly-wheels fifty-six feet in circumference. The cylinder is twenty-six by twenty-eight inches, and the arm attached to the fly-wheel, weighs eight thousand six hundred pounds.

The engine is set in the centre of one end of the building, and rests on eleven stones laid in hydraulic cement. These stones weigh eighteen

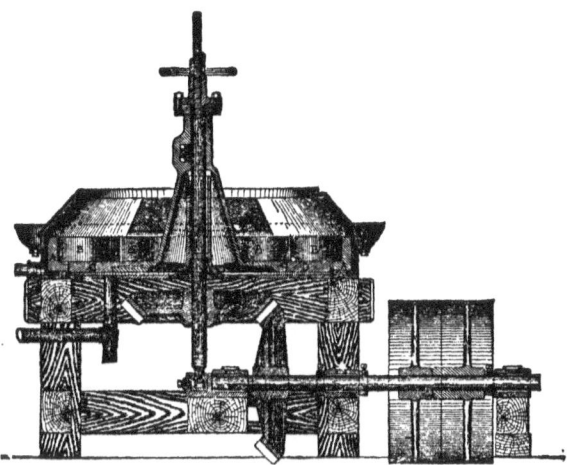

THE HOWLAND PULVERIZER.

The ore is fed through the opening in the top of the bonnet-casting; immediately on falling upon the revolving disc-plate, it is carried outward by centrifugal force to the rings or rolls, and when pulverized fine enough, is ejected through the screens to a circular trough conveying it to copper plates for amalgamation, or run into tanks for settling. The machine is constructed also for dry pulverizing; it will pulverize about one-fourth less dry than wet. The frame for the machine is made of southern pine timber, is mortised and tenoned throughout and held by strong joint-bolts. Each machine is put together at the works, and finished, marked and taken down for shipment. An automatic feeder is provided especially for this machine, that will feed the ore continuously. A rock breaker is provided also which is particularly adapted to breaking the ore to the proper size for this machine, which must not be larger than one inch. This machine will pulverize wet, hard quartz rock to a fineness that will pass through a 40-mesh screen, *one ton per hour*, and will pulverize dry to pass through a 60-mesh screen, half to three-quarters of a ton per hour.

INGERSOLL ROCK DRILL.

The cut illustrates the adjustable column for supporting the drill when working in tunnels and drifts. Compressed air or steam, as may be most convenient, are the motors employed. One small drill of this character is guaranteed to do more work on a heading than nine men could do, provided so many could find room to work there at once.

thousand pounds each. Two line shafts on each side of the building have Walden & Mason's patent friction pulleys to drive the stamps.

The great benefit of the patent friction pulleys is, that each ten stamps of the mill can be stopped by means of friction, the shoes being thrown in or out of gear by a lever without interfering with the motion of the mill.

Three of Blake's largest size rock breakers are set in the top of the mill, so arranged that the rock when partially crushed passes to twenty-four of Hendy's patent self-feeders, placed in the rear of the stamps. The mortars are lined with copper and have improved screens. The stamps weigh eight hundred pounds each. The Hendy concentrator is a new feature in this part of the country. There are twenty-four of these placed at the end of each plate, so that, as a safe-guard, should any gold escape the battery or plate, it is saved by the concentrator. The tailings coming from the plates are run into the concentrator, which acts upon the same principle as we would pan out dirt by hand. This is only needed when the pyrites or sulphurets of iron contain gold, as quicksilver will not act upon pyrites of iron. The machinery weighs over one million pounds, and the belting used weighs over three tons. The mill cost two hundred thousand dollars.

The hoisting machinery consists of two twelve by twenty-four engines of seventy-five horse-power, which are capable of raising two tons four hundred feet per minute ; two reels for hoisting, supplied with a steel wire cable which will sink the shaft to a depth of one thousand feet.

This cable is a most powerful and costly one, being two and three-fourths of an inch in diameter, weighing one pound and a quarter to the foot. Added to this is a six-inch drawing and lifting pump, with five-feet stroke, having a capacity of five thousand gallons per hour. The entire enterprise is one of *immense* proportions, and one that will be a profitable monument to the nerve and money of San Francisco capitalists. The Homestake Company has paid regular dividends of thirty cents per share on one hundred thousand shares, aggregating thirty thousand dollars monthly, for sixteen successive months up to April 1st, 1880, making a sum total of four hundred and eighty thousand dollars, which the mine has paid its shareholders in a year and four months.

The Caledonia Mine

is opened by a tunnel which runs into the hill nearly at right angles with the ledge, which is reached at a distance of three hundred feet. At this point an immense chamber of ore has been excavated, and a shaft eighty feet deep connects this chamber with the surface. The main tunnel is being now driven along the foot-wall. It is about six hundred feet long, and at the north end a drift forty-five feet from the foot-wall to crosscut the vein, shows the face and side in solid ore, and for the whole six hundred feet of this tunnel, the top, bottom and side show a continuous body of ore, the extent of which can be determined only by cross-cuts. One hundred feet below, and further down the hill, another tunnel has been started for the purpose of tapping the vein immediately under the large chamber, where connections will be made by means of a winze now being sunk. The lower tunnel is in about five hundred feet and has struck the ledge, a vein ten to twelve feet wide, which shows specimens of free gold. The company have a twenty-stamp mill which crushes about forty tons of ore per day, which averages nine dollars and fifty cents per ton. About thirty men are at work in the mine, but many more could be advantageously worked if necessary. The company expect to construct an eighty-stamp mill this season (1880) and assessments have been levied for that purpose.

The "Clara Consolidated," a mine lying parallel with the Caledonia, has been listed on the San Francisco stock board. It is opened by a tunnel, which is being driven into the hill as rapidly as possible.

"The Golden Terra" Mine

is a well-developed property. Over one thousand two hundred feet of tunnels and cross-cuts show up an enormous amount of ore in sight. One drift has a cross-cut one hundred and ten feet in length, which is in ore the entire distance. Another drift, running parallel with this main drift, shows the same character of ore. The main lower tunnel is in three hundred and sixty feet, and will run parallel to the upper tunnel, and be connected with it by winzes at convenient distances. Four veins have been cut in running the lower drift, one showing forty feet of ore, very similar to that on the lower level of the Father Desmet. This assays seventeen dollars per ton. At the surface the dirt from the grass roots down is being milled. A large breast of ore, two hundred feet wide, is exposed, which will be quarried down, and, by means of a chute into the tunnel below, transported by cars to the mill. A single blast has thrown down several tons of ore. No ore has been milled that did not yield from ten to twelve dollars per ton. The company are running a thirty-stamp mill, and are building another sixty-stamp mill. About fifty men are worked in the mine.

We would gladly, in this connection, devote more time to the description of the many rich mines in this vicinity, which for lack of space we are obliged to omit. We have, however, described some of the large ones, those most developed, which will give an idea of the vast deposits of ore which these Black Hills mines contain.

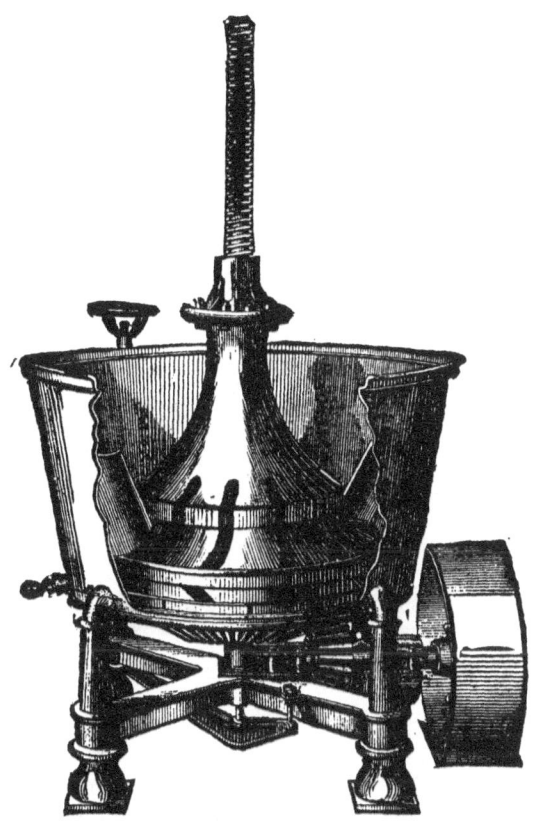

EXCELSIOR GRINDING AND AMALGAMATING PAN.

The Pan has the conoidal form, the centre rising as high as the rim, and molded so that its vertical section forms the tractory curve, or curve of equal wear, securing perfect uniformity in the wearing of shoes on the muller and the dies. Its mechanical construction as to simplicity, weight, strength, convenience of working, cleaning up and cheapness, it is claimed, is unequaled by any other grinder and amalgamator.

CHAPTER X.

BLACK HILLS SILVER MINES—FIRST DISCOVERY—THE FLORENCE MINE AND MILL—THE CORA MINE—THE BALD MOUNTAIN DISTRICT—THE ROCKERVILLE PLACER MINES, ETC.

THE silver interests of the Black Hills, though yet in their infancy, should not be overlooked. They give promise of great things for the future, and the discoveries now being made are of that importance which make this interest second to none in the Black Hills.

The Bear Butte silver region and its city of Galena, a thrifty, growing town, will, in time, become a region of great importance.

The First Discovery of Silver

in the Black Hills was made in the Bear Butte District, some ten miles south-east of Deadwood, in 1877, by mere accident. A miner was prospecting for gold on a neighboring hillside, and came upon a large body of ore, which contained about twenty per cent. of lead and a fair proportion of silver. About the same time another miner, who was sluicing in the gulch, found among his gold dust, one night, a nugget of almost pure silver, which had evidently been washed down from some vein in the mountains above. These circumstances led to a search for that metal, and those who could content themselves with the less valuable silver ore found it in all directions. One of these silver ledges has been traced for five miles, and openings made upon the lead at various points nearly the whole distance.

The Florence Silver Mine

was located in 1877, and was purchased in 1879 by some St. Paul capitalists, who have erected a fine silver mill for the reduction of their ores. The mine is well developed by tunnels and drifts, and has in sight a large amount of ore, which averages from thirty to one hundred and fifty ounces of silver per ton, or about that many dollars. The mill reduces about twenty-five tons of ore per day, and the prospects of the company seem flattering. There is at Galena a small custom smelter, and other silver mines of great promise. The Florence mill is turning out pure silver bullion at the rate of about twenty-five thousand dollars per month, and has as fine machinery as can be found in the country. Each alternate day a brick is made, that weighs one thousand to one thousand three hundred ounces of silver. The mill is a twenty-stamp dry crusher, with two Bruckner cylinders, for roasting the ore.

The Florence Mill.

The mill is at the base of the hill, and on a line with the mouth of the tunnel. By means of a tramway and chute the ore will be dumped into the ore-room, where it is partially crushed by a Blake rock-breaker, from whence it goes to a drying cylinder and is thoroughly dried, and thence to the self-feeders, where it is fed into the stamps, and thence to the Bruckner cylinders, where the silver and lead, by this roasting process, is oxidized, the baser metals of arsenic, sulphur and antimony being separated, the silver and lead dust thence passing to mortars and concentrating pans. The mill cost about eighty thousand dollars.

In the Galena silver region the stratification of the county rock lies horizontal, the formation being granite and porphyry, capped with limestone and quartzite.

Most of the work has been done in the quartzite, where rich chlorides, sulphurets and carbonates of silver outcropped at the surface, in the form of "chutes" or "chimneys." These have been followed by tunnels, whereas, had the quartzite been pierced by shafts until the primitive formation of granite was reached, it is probable that a "deposit" or a contact vein of ore would have been exposed. However, there are a number of true fissure veins in the district with well-defined walls.

Mr. Robert Floorman, the superintendent, and we believe, discoverer of the Florence mine, has driven a tunnel into the mountain a depth of over six hundred feet, and is not far from the granite formation of the main hill. The indications near the end of the tunnel point to the theory of a large deposit of ore. The numerous "pockets," filled with chloride and sulphurets of silver, are more frequent and larger, while the pay streak is taking a downward dip toward the centre of the hill, where it will probably take the granite for its foot-wall. The superintendent had about one thousand tons of ore on the dump, graded as first, second and third class. The first class assays about one thousand dollars, and the third-class ore about thirty dollars per ton. Another silver mill, called

"McDonald's Furnace,"

is in the district, and is pronounced a good one, and is a custom smelter. The most prominently-developed mines, besides the Florence, are the "El Refugio," the "Meritt, No. 1," "Meritt, No. 2," the "General Custer," "Spotted Tail," "Buckeye," "Sitting Bull" and "Crestline."

What is needed in the district is more custom-smelting works, or a sampling and ore-buying company, so that the miners can sell their ore as fast as extracted. Many of the miners have several tons of high-grade ore on their dumps, on which they would be unable to get a sack of flour.

The Cora Mine,

at Galena, in which a recent rich strike was made, is on the same vein as "Meritt Nos. 1 and 2," "R.

B. Hayes," "Sitting Bull," and others. This vein is opened every three or four hundred feet for three miles. Assays have been made from the Cora, showing an average of twelve thousand dollars to the ton, in one instance. The mine was located in 1876, and considerable work was done at various points, showing large bodies of Galena, or chiefly lead ore; but it was only lately, while a new shaft was being sunk, that anything like the value of the property was known to its owners. Then near the surface they encountered a body of decomposed Galena, mingled with black sulphate, in which native silver could readily be seen.

Reports of the discovery spread like wildfire throughout that and adjoining camps, and the greatest excitement prevailed. The lucky owners of this Bonanza are W. W. Andrews, P. Donovan and J. McNamee. The vein proper has a width of about four feet, but the rich carbonate streak above alluded to is only about eighteen inches across. It is impossible, at this early date, to give any correct estimate as to the future value of this mine. It would be the exception, and not the rule, for the mine to produce ore of such great richness in any considerable quantities; and although the mine may be in fact a regular "Bonanza," it is more than probable that this rich pocket of ore will soon be exhausted.

There is also a gold district in the vicinity of Galena City, in which much activity is being shown,

The "Boss Tweed" mine is down one hundred and twenty-five feet in ore, which assays well in gold. The "Golden Prize" mine, in the same

The Bald Mountain Silver District.

district, produces some wonderfully rich gold specimens, and in Strawberry Gulch, six miles out toward Deadwood, there are several promising mines. The Keystone, of this district, has produced ore assaying one hundred dollars per ton, has encountered a large body of ore; much of the ore, however, is of low grade. From eight pounds of rock from this mine one of the hands is said to have taken out two dollars' worth of wire gold with his fingers from a piece of ore.

The "Sunday Mine,"

of the same district, has opened up some fine specimens of gold-bearing quartz at a depth of one hundred and twenty feet. In cross-cutting seventy-five feet, twenty feet of ore were found to be quite rich. A forty-stamp mill has been erected at this mine.

The Bald Mountain Silver District, south-west of Deadwood, is also attracting much attention. It would not be strange if the silver interests of the Black Hills, which are yet in their infancy, should prove of equal importance to her gold interests, and perhaps of even more permanent character.

The points which especially seem to give the mines of the Black Hills advantages over those of some other localities, are:

1st. The nearness of the pay-ledges to the surface of the ground, it not being necessary to sink very deep shafts, or long and expensive tunnels to reach them.

2d. The great magnitude or size of the veins or ore bodies, and the consequent cheapness with which they can be mined.

3d. The freedom of the ores from base or refractory metals, which admits of their being milled by the simplest processes; or, in other words, their free-milling qualities.

4th. Their accessibility, or the ease with which the mines can be reached with good roads.

5th. The abundance of fuel, timber and water for mining and milling purposes.

In many cases the veins of ore lie so near the surface that they have been opened by merely stripping off the thin layer of soil over them, exposing the ore which has been quarried out as from a great quarry, at small expense, and has paid for milling at the very grass-roots.

These ledges of reddish-brown, iron-stained quartz, from fifty to two hundred feet wide, lying upon their edges between walls of slate, at an angle of about fifty degrees, are *all ore* from wall to wall, and is sent to the mill, in many cases, without selecting. The ore is easily blasted down in large quantities, and frequently the cost of shoveling it up into cars is saved by shooting it through a winze into cars in a tunnel below, where a track is lain to conduct the cars into the tops of the stamp-mills, where it is dumped into bins, and almost by its own gravity is fed automatically into the crushers and stamps below, saving a vast expense in labor. In this way the expense of mining and milling has been reduced to a trifle, as compared with the usual cost of handling ores.

This explains how it is possible to mine and mill ore at an expense of only from two to five dollars per ton in the Black Hills, and how ores, only milling eight to ten dollars per ton, can be made to pay a handsome profit.

Ores can be reduced much more economically in large mills than in small ones. It requires about as many hands and superintendents to operate a twenty-stamp mill as it does one of a hundred stamps, while the latter will crush five times as much ore. Hence the companies here have taken advantage of this fact, and have erected some of the largest stamp-mills in the world.

CHAPTER XI.

TAKING THE BULLION AWAY—ROAD AGENTS.

The Black Hills.

ALTHOUGH at present there is as much safety and security to person and property in the Black Hills as in any country on the frontier, yet there was a time when, tempted by the heavy shipments of gold dust and bullion *via* the stage lines, highway robberies by armed men were frequent. Road agents, as they were called, stopped nearly every unguarded coach which passed between Cheyenne and Deadwood; and it became necessary to have every treasure coach strongly guarded.

Express rates on bullion and gold dust consequently became very high. Most of the bullion produced in the Black Hills is shipped to New York *via* Sidney and the Union Pacific Railroad, and the charges are one and one-half per cent., equal to fifteen dollars on every one thousand dollars. Of this amount the stage companies receive one per cent., and the railroad company the balance. For a long time no treasure has been shipped except by special coach, sent twice a month, guarded by eight armed men, besides two others who ride at a distance ahead of the coach, and two behind also. These men never rest or sleep until they have reached their destination at Sidney. The coaches carry nearly two hundred thousand dollars each trip.

Stage Robberies.

Only one treasure coach has ever been robbed, and that was only guarded by two messengers, and contained but twenty-five thousand dollars. This amount was taken, but eighteen thousand dollars of it was afterward recovered and two of the robbers captured.

During the early spring, in 1878, several coaches on both the Cheyenne and Sidney route were robbed in quick succession.

On the 26th of June, 1878, the treasure coach above mentioned was attacked near the Cheyenne River. After the exchange of a few shots, which wounded the driver, the highwaymen won a complete victory over the messengers. The passengers were all ordered out and placed in line alongside of the road, and compelled to hold up their hands, with the muzzles of cocked revolvers pointed at their faces, while one of the robbers searched their pockets and baggage for valuables. Failing to unlock the treasure-box, and not being able to break it open with hatchets, it was proposed to abandon it, until one hit upon the plan of blowing it open with gunpowder. This happy thought was quickly put into execution, and in less time than it takes to write it, the box was blown open, and the plethoric purses of gold dust extracted, after which the coach was allowed to pass on.

On the 1st of July, in the same year, while the Cheyenne coach was passing near the same place, in the night, two men raised up suddenly at the side of the coach and demanded it to halt. One ran ahead and stopped the leaders, and ordered the passengers to alight. All obeyed. After reaching the ground the men stood in line, with one road agent behind as a guard, with a gun and two revolvers. The other robber began at the opposite end of the line, from which a man named Flynn stood, and commenced going through the passengers. When he had reached the man standing second from him, Flynn whispered to his nearest comrade to look after the unarmed robber and he would attend to the guard behind, and suiting the action to the word, drew a pistol and fired, the robber dropped to the ground, and almost immediately Flynn received a shot in the face, after which he ran around behind the coach and hid in the bushes, the robber firing at him all the time, and he returning the shots. The driver whipped up and drove on, but soon returned with two men and took aboard the passengers, two of whom were wounded—the robbers escaped.

On the 25th of the same month, before daybreak, the stage was again stopped near Hat Creek by six armed men, who were on foot and masked. Finding there was but one passenger on board, and that he was a preacher, they forebore molesting him, but turned their attention to the mail-sacks, which they cut open and robbed of registered letters and other valuable matter. They also broke open the treasure-boxes, but found nothing.

They were not at all excited over their work, which occupied them a full half hour, and as soon as they were through they ordered the driver to go on. About the 13th of September, near the same place, the coach was fired into and a man with a mask, from a distance, ordered the driver to have the passengers get out. There were two, a lady and gentleman, who very willingly complied with the request, and the man was ordered to hold up his hands, when his arms were securely tied behind him with a buckskin string. The captain of the gang then told him he had taken a fancy to his ulster overcoat, and must have it, so the pinioned arms were relieved and the coat removed. His pockets were next searched, although he assured the robber that it was a foregone conclusion he was broke. Five dollars were, however, obtained. Two pocket-books were examined, but nothing contraband being discovered, were handed back. The lady was next called upon, but responded very feebly—only one dollar and a quarter.

The investigation ended, both were ordered to forward march down the road, when, in a few minutes, about two hundred yards away, a sight for an artist met their gaze—four men and one woman, shivering in the cold; the hands of the men tied behind them, and crouching under the ghostly light of the lamp of the south-bound coach and the awe-inspiring glances of a double-barreled shot-gun.

The twain were ordered to stand a few feet away, and the rifling of the pockets of the south-bound passengers began. The messenger was the first to be overhauled, and a shot-gun, watch and chain, valued at one hundred and fifty dollars, and two pistols, overcoat and fifteen dollars in cash were taken. Another man gave up ten dollars, but the bandits, dissatisfied, wanted his boots, and asked what size they were. He replied by sitting down on the grass and holding his foot aloft, while the captain placed his against it, but found that a difference of several sizes existed. Another passenger was rifled, but did not return a satisfactory amount, as the petulant looks and grumbling tones of the agents evinced. Lastly, they approached a lady of the south-bound coach and took from her fingers several valuable rings, including her wedding-ring. For this she plead and begged, and upon her knees asked that the little trinket be returned to her; but supplication was unavailing, the ruffian gruffly informed her that a wedding-ring was as good for him as her. Before leaving, the driver yielded to the solicitations of the gang, and gave up six dollars in money and a silver watch. The treasure-boxes of both coaches were broken open, but nothing but envelopes found.

The work of plundering being finished, the captain ordered a subordinate to level his gun upon the party, while he went around the bluff and got the horses, which he hitched near by. Two robbers covered the passengers with their rifles, while a third mounted his horse, returned and leveled his rifle until the second could secure his animal, and so on until all were mounted; they then bade the passengers good-night and rode away. When about one hundred yards distant, the captain discharged his weapon in the air, probably as a signal. One of the men's arms were released before the ruffians left, and he and the ladies untied the other passengers. Several of the party congratulated themselves that a more thorough search was not made, as one had ten dollars under the lining of his hat, another a substantial roll concealed in his coat-sleeve, and the lady one hundred and ninety dollars within her stocking, under her foot.

Two days later, the villains again attacked and robbed two coaches in the same locality. The north-bound coach left Hat Creek at nine o'clock in the evening, and had been out about two hours when, in the middle of a short, steep hill, the not-unlooked-for cry of "Halt" came. Below the coach and lying upon their faces were four men, pointing each a business-looking rifle; one directing his to the lead horses, another to the driver, while the other two were trained upon the coach. The spokesman asked the driver, "How many passengers have you?" receiving the answer, "Two, a lady and gentleman." "Get out and hold up your hands." The command was obeyed by the gentleman, but the lady was allowed to retain her seat. The gentleman was approached and asked, "How much money have you?" He replied, "Thirty dollars." The highwayman said he would take one-half of it, but by advice of his comrades compromised with ten dollars, giving that back when he ascertained that his victim was a laboring man, with the words, "We don't want to molest the passengers, but must have the treasure." They next asked for eatables, and the driver produced a box of fruit in reply, which was broken open and a good portion devoured, the ruffians offering the passengers a share.

Robbing the Mails.

The mail-sacks were handed down, and one cut open and one found unlocked. Their contents were poured out upon the ground, sorted over and returned to the pouches. The gentleman's trunk was taken off the rack, but not broken open. The driver was then ordered to go on, and meeting the south-bound coach in about two hours, he informed the outriders of the adventure of his coach a few hours before. When the south-bound coach reached the same spot it was also stopped. There were two passengers in this coach and two messengers following about two hundred yards behind on horseback. As soon as the coach was halted, the messengers dismounted and approached within fifteen steps of the thieves, one of whom called to them, "Halt," accompanying his command with a shot. One of the messengers returned the shot, killing the thief who had fired. The remaining highwaymen at this moment began firing at the messengers and retreating toward the gulch close by, to which point the messengers could not follow. The coach, meantime, had been ordered to go ahead by the thieves, who had succeeded in robbing one passenger and in securing the mail-sacks before the fight began. The messengers held their ground for half an hour after the firing ceased, but could not reach the place where the mail-sacks and the dead robber lay, and not being in sufficient force to dislodge the thieves, they mounted their horses and joined the coach.

Besides the robber who was shot dead, it is known that two others were badly wounded. Both the messengers escaped unhurt. It was learned from the passengers, that the reason the money was returned to the gentleman, was because upon being asked by the captain what his business was, he relied, "A working man."

At this period nearly every outgoing coach was

stopped, and travelers expected to be halted, and were obliged to adopt the plan of carrying very little money or valuables with them. The road agents found on the person of a passenger one night a cheap watch, and besides refusing to take it, ridiculed him for carrying such a timepiece. At one time the robbers got only fourteen dollars from six passengers, and remarked that it was d——d strange that folks traveled with so little money about them. One of the passengers replied, "That none but fools would carry money, knowing they were sure to lose it."

Hence it soon became unprofitable business for the road agents, and it soon became also accompanied with too great risk to pay them to follow it. The messengers became more bold, and several of the highwaymen were killed, others were captured by the sheriff and posse, so that finally the business was entirely broken up and no further molestation occurred.

CHAPTER XII.

COAL, OIL, SALT AND AGRICULTURAL RESOURCES, RAILROADS, ETC.

The Black Hills.

ON the 5th of January, 1877, while hunting north of the Redwater River, James Brewer discovered a bank of coal, at a point where it cropped out of the mountain. Subsequently other discoveries of coal were made at various points in the Black Hills country, and at Sturgis City, and also near Rapid City small veins were found, but nothing of very great importance except that in the Redwater region.

This coal bank was named by Brewer the Blossburg Coal District, and lies about nineteen miles north of Spearfish City, and thirty-one miles north-west of Deadwood. On the 1st of June, 1878, four companies were organized, locating two thousand five hundred and sixty acres of land. During the same month the four companies consolidated, under the management known as the "Black Hills Consolidated Coal Mining and Fire-Clay Manufacturing Company," Captain James Christie, General Manager, and Robert Chew, Treasurer. Under this management they now have a tunnel in two hundred and fifty feet, with rooms on each side, and will soon have their mine in shape to get out any amount of coal that may be required for consumption in the Black Hills. The vein is five and a half feet thick, with indications of widening. The coal is bituminous, a superior article, and easily mined, and there is every reason to believe that it will make good coke, which the company propose to test. They have prospected their entire body of land, and find the same vein of coal extending almost through their entire location. There has been a great deal of prospecting south of Hay Creek, near Redwater River, and the indications of coal are found, but the vein is only from eight to fourteen inches thick, and has what they call "horses" extending through it besides, which would not pay to work. There is a good road from Spearfish City to the Blossburg coal bank, and the company propose to build a railroad from Deadwood to their mines as soon as practicable. There is little doubt but that the Black Hills coal deposits are sufficient to abundantly supply the country with fuel when they shall become properly developed.

Oil.

Another one of the rich resources of the Black Hills which cannot be overlooked is the petroleum wells.

Strange as it may seem, a hunter was destined to discover these as well as the coal banks. A man named Taylor, while having his Christmas hunt on the splendid elk and deer ranges, near Jenney's stockade, during the closing week of 1877, one day camped by the side of a spring which he found utterly worthless to quench his thirst, and, as he afterward expressed it, "the water was only fit for shoe grease." News travels very quickly, sometimes, even in the rough mountain passes of the Rocky Mountains, and it was not long until his disagreeable experience with the spring became the subject of bar-room jests among the mining camps. So, early in 1878, a small party of Pennsylvanians, who were largely interested in quartz and placer mines at Deadwood, quietly organized a prospecting expedition, braved storm and other hardships, and looked the ground over thoroughly. The experienced Oil City men were thoroughly at home among the immense beds of oil-bearing shale, the croppings of slate and sandstone, and the mires of seeping petroleum. Four of them located one hundred and sixty acres each, cornering at the largest spring, and others staked off equally good claims.

Before leaving they erected cabins, made a thorough examination of the surroundings, and filled quite a number of bottles with samples of the oil. Some of these samples were at once tested by machinists and engineers, at Deadwood and in the East, and pronounced by all to be equal to the oils of Pennsylvania. Pennsylvania experts declared this oil to be almost totally free from grit or other impurities, and of very heavy "body," possessing a gravity of from thirty to thirty-three.

This Black Hills product has since been dipped up by the barrelful from the largest spring, where it escapes at the rate of over a barrel per day, and is used exclusively for lubricating purposes by several quartz mills in the Black Hills.

Black Hills Petroleum.

After the pioneering had been done, of course there were plenty to turn their backs upon the glittering treasure at Deadwood and stampede to the less romantic petroleum fields. In a few days over one hundred locations had been made, and a miniature oil city created, by the dozens of log cabins which immediately sprang up.

The fields are undoubtedly very extensive, having already been traced for a distance of twenty-five miles in one direction, and possessing a width which is yet unknown. They lie along the Cheyenne and Black Hills stage road, about six miles from Jenney's stockade and fifty miles southwest of Deadwood, being about two hundred miles north of the Union Pacific Railroad. They are, therefore, easily accessible, and will doubtless soon be crossed by the iron rails.

The valley of the South Cheyenne River, about three miles distant, and the smaller valleys at the base of the hills, have fine areas of fertile agricultural lands, while the hills adjacent are covered with heavy growths of pine and other kinds of timber. We are informed that machinery has been taken there for boring, and that numerous wells will undoubtedly soon be spouting oil. Common lubricating oil sells in Deadwood for twenty-five to thirty dollars per barrel, and ordinary illuminating oil, which in the East sells for fifteen to twenty cents per gallon, costs in the Black Hills from sixty to ninety cents. The oil taken to Deadwood, from Jenney's stockade, has sold to the mills for lubricating purposes at from one dollar to one dollar and twenty-five cents per gallon.

While the market afforded by the wonderfully-increasing population and enterprise of the Black Hills would be quite an item, the fact should not be overlooked that there are no oil wells successfully worked in all the trans-Missouri region, and that the consumption of lubricating and illuminating oils in this broad portion of Uncle Sam's domain, runs up in value to millions of dollars annually. If present prospects are one-half fulfilled, the export of oil from the Black Hills' wells will, inside of a few years, add largely to the general prosperity of the region.

Black Hills Salt Springs.

Worthy of consideration in connection with the great treasure-vaults of silver and gold, and the deposits of coal, lead, mica, quicksilver and petroleum, and the generous breadths of fertile agricultural and pastoral lands, and splendid forests, are the Black Hills salt springs and wells.

In the vicinity of the oil wells, near Jenney's stockade, are several large and valuable salt springs. The water, which is thrown off in great quantities, is found to hold in solution one pound of very fine salt to one gallon of liquid.

Near the base of Inyan Kara Peak, is Salt Creek, which may or may not be the stream which certain political parties so often navigate. So strongly are the waters there impregnated with salt, that crusts of this staple, two and three inches thick, and almost chemically pure, are formed on the gravel-beds which have been recently overflowed.

Black Hills Salt Works.

Thirty-five miles south of Deadwood there is another cluster of salt springs, forming a miniature salt river, on whose banks are located the salt works of Henderson & Co. These works produce a beautifully white and superior article of salt, by evaporation, which obtains ready sale in the Black Hills cities and mining camps. There are six large iron evaporators in use here. An invoice of four thousand pounds which was sent to Deadwood, was accompanied by one of the owners, who declared that enough salt could be made from his springs to supply the whole West.

Railroads.

There is much talk in the Black Hills about railroads, and the desire of the people is strong to get an outlet for their resources and cheaper transportation for goods and merchandise. There is little doubt but in the course of time their hopes will be realized, as it seems hardly probable that a country so rich in resources as the Black Hills will be long left out in the cold by capitalists, especially since capital has invested such vast amounts in Black Hills mines.

A company was organized in Cheyenne more than a year ago for the purpose of constructing a line from the Union Pacific at that point to Deadwood. It was rumored that Jay Gould would push the enterprise to completion, making it a branch of the Union Pacific. A branch from the Northern Pacific, south from Bismark, has also been contemplated; other projects have also been mentioned. The Chicago, Milwaukee and St. Paul Railroad Company, and also the Chicago and North-western Company, have each a line completed nearly to the Missouri River, and it is thought both may eventually extend their lines to the Black Hills. But how soon any of these lines may be pushed to Deadwood, or if any of them will be ever completed, is at present guesswork.

Agricultural Resources of the Hills.

When the high prices and ready market for all kinds of grain and produce is considered, it must be admitted that the Black Hills offer no trifling inducements to the agriculturist and stock raiser. The Black Hills contain districts of splendid farming lands, the principal part of them being the valleys of the Spearfish and Redwater streams on the north, and the Rapid and French Creek valleys on the south and east.

Those of the Spearfish and Redwater embrace an area of not less than one hundred and fifty square miles of good land; enough for six hundred farms of one hundred and sixty acres each. Of this area only about two hundred locations have been made, and only about fifty of these have been improved by cultivation. Crops of all kinds can and have been successfully raised here. One Spearfish farm netted its owners, in 1877, fifteen thousand dollars, from potatoes alone, after which the owners sold their improvements for three thousand dollars. Irrigation is necessary to successful farming here, but the supply of water is ample to irrigate all the lands in these valleys. The cultivatable extent of the Rapid valley is about two hundred square miles; or enough for eight hundred quarter-section farms. Probably about one hundred to one hundred and fifty locations have been made, leaving room for nearly seven hundred more. These lands can be easily irrigated from the Rapid Creek, and the supply of water is unfailing.

CHAPTER XIII.

DISCOVERY OF GOLD IN THE CENTENNIAL STATE.

Colorado and the Gunnison.

FOR ten years after the wonderful discoveries of 1848-49 in California, a stream of immigration had been pouring across the continent, over a route a little to the north and in plain view of the snow-clad summits and glistening peaks, beneath which and in whose bosom lay treasures of gold, only equaled by the rich finds on the Pacific Coast. Little the traveler dreamed that here, half-way on his long journey, almost at his feet, lay the "Golden Fleece," of which he was in search.

The "crisis," or hard times of 1857, was the incentive which drove some dissatisfied adventurers from the East, among whom were a few Georgia and California miners, to search for new fields for discovery and adventure. Here was a vast unknown region suitable for their purpose. It had then not even a name. Lying on both sides of the Rockies, it might properly be called a saddle on the "back-bone of the continent"—that lying on the eastern slope was considered as a part of Kansas, and that portion west of the range a part of Utah Territory.

In 1858 this little band of explorers, headed by Greene and Russell, passed up the Platte River to the foot-hills of the Rocky Mountains, and a few of the experienced miners began to prospect for gold. Panning out some of the earth along the streams they found colors; this led to a more extended search, and resulted in the finding of small quantities of the metal and a few gulches of considerable richness.

The Pike's Peak Gold Excitement.

This was enough to start the rumor of *gold* and to attract a considerable population to the district before the autumn. This section became known as the Pike's Peak gold region.

In 1805, Lieutenant Pike had visited the region with a few soldiers, and passing on down to the Rio Grande River, he was captured by the Spanish military forces stationed in that vicinity, who at that time claimed the country; but he was afterward released. From this fact a prominent peak was named Pike's Peak, in his honor.

In the fall of 1858 the pioneer gold hunters founded several towns, among which were Denver, Auroria, Boulder and Fountain City. With the spring of 1859 there came a rush to the new-found gold diggings, such as can only be compared to the one of ten years previous to California. The road across the plain during the spring months was one continuous line of vehicles and prairie "schooners," a caravan of excited pilgrims, all bound for the land of promise.

"Pike's Peak or Bust."

It is said that an inscription on one of those canvas-covered equipages at that time read, "Pike's Peak or Bust," and we presume, if the unwritten history of that outfit could be known, the probabilities are that it "*busted;*" for it is a fact that about nine out of every ten of the stampeders to the region at this time met with disaster or misfortune. Probably fifty thousand men, more or less, aided in swelling the population of the then unnamed Colorado during that year. Of course, there were not opportunities for such a vast influx of people to find profitable employment in the limited extent of country then known to contain gold. As is always the case, many were young men, ignorant and inexperienced in mining, and altogether unfit for the trials and hardships of a new country, far beyond the limits of civilization. Therefore, many returned ere the summer was gone, disheartened and discourged.

As the population increased, miners began to extend their explorations into more distant localities. No mountain was so steep and no cañon so deep and rugged as to shut out the ever-resolute seeker for gold.

About this time, quartz veins of gold-bearing ores began to be discovered and located. Among

SUMMIT OF PIKE'S PEAK, COLORADO.

the most prominent quartz districts at this time was that in the vicinity of the present towns of Central City and Black Hawk. John H. Gregory discovered a great gold-bearing vein, which has since been known by his name, and which is said to have yielded more money than any other fissure in Colorado. Thus began a new industry; stampedes were in order all over the State, to points wherever a new district was formed or quartz mines discovered, and locations of mines were plentiful.

The first stamp-mill was brought into the Gregory district in 1859, and the first newspaper was established in Denver, in April of the same year, and called the *Rocky Mountain News*. During this time the placer mines, though not extraordinary rich, were yielding bountifully. Some gulches were producing wonderful amounts of gold.

The yield of gold in Colorado, for the year 1859, was not far from five hundred thousand dollars, and that of the following year (1860) was over three millions. In the spring of 1860, mining was very active, new-comers to the gold-fields were still very plentiful, Denver and other towns were growing very fast. Over fifty quartz-mills were brought into the country and had commenced pounding out gold. About thirty Mexican arrastras were also in use grinding out the gold ores of the section. During the summer, some miners in search of new diggings entered California

Gulch, now in Lake County, on the banks of which the city of Leadville is situated, and found gold there. Simultaneously other new districts were discovered, and mining was exceedingly lively that summer.

Gregory and Russel Gulches, now in Gilpin County, yielded abundantly. California and Georgia Gulches produced an astonishing amount of gold dust; some diggings having, it is said, averaged a pound of gold per day to the man, and frequently an oyster can was filled with gold in a single day from one claim. Spring Gulch, at Central City, also produced a considerable amount of dust. These gulch and placer mines in four years after their discovery had, it is estimated, yielded over fifteen millions of dollars in gold. The gold quartz mines at this time had also begun to produce largely. Silver had not yet been discovered.

A Bit of History.

In February, 1861, Congress organized the Territory of Colorado, and fixed its boundaries the same as the present State. Previous attempts at organization had failed, and the country had only succeeded, through a representative sent to the Kansas Legislature, late in 1858, in getting the County of Arapahoe organized, embracing all of western Kansas, reaching to the Rocky Mountains. The first Governor appointed was William Gilpin, who reached Denver in May, 1861. The new Territory embraced an area of one hundred and four thousand five hundred square miles, or more than any other State except Texas and California. A census taken at this time showed a population of about twenty-five thousand, of whom only four thousand were females. In 1862 and 1863, large numbers of miners left the failing placer mines, and sought new and distant camps. About this time, Montana and Idaho came to the front as gold producers, and the fame of Alder Gulch was spread far and wide.

Still the mines of Colorado continued to produce largely, and up to the present time (1880), have produced, in gold and silver, not far from one hundred millions of dollars. The year 1879, alone, produced a little less than twenty millions in gold and silver bullion, and in the twenty years in which mining has been actively carried on in Colorado, an average of five millions per annum has been reached.

The following tables will show the yield of gold, silver and copper, etc., for the various years since 1858.

Year.	Coin Value*	Year.	Coin Value.*
1859	$500,000	1865	$2,525,000
1860	3,250,000	1866	1,575,000
1861	3,250,000	1867	1,750,000
1862	3,400,000	1868	2,000,706
1863	3,400,000	1869	2,482,375
1864	3,350,000		

* Some of these estimates given by "Fosset's Colorado"—other authors differ.

EXCESS OF SILVER.

Year.	Gold.	Silver.	Copper	Lead.	Total.
Previous to 1870. Total	$27,213,081	$330,000	$40,000		
1870	2,000,000	650,000	20,000		
1871	2,000,000	1,029,046	30,000		
1872	1,725,000	2,015,000	45,000	$5,000	$3,790,000
1873	1,750,000	2,185,000	65,000	28,000	4,028,000
1874	2,002,487	3,096,023	90,197	73,676	5,262,383
1875	2,161,475	3,122,912	90,000	60,000	5,434,387
1876	2,726,315	3,315,592	70,000	80,000	6,191,907
1877	3,148,707	3,726,379	93,796	247,400	7,216,283
1878	3,490,384	6,341,807	89,000	636,924	10,558,116

For the year 1879, the total production nearly doubled. Various estimates differ greatly from different sources, but we give below one which is generally conceded to be correct.

Production of Colorado, 1879.

Lake County	$11,477,046	Gunnison	$300,000
Gilpin "	2,608,055	Summit	295,717
Clear Creek	1,912,410	Chaffie	71,240
Boulder	800,000	San Juan	483,500
Custer	720,000		
Park	434,449	Total	$19,102,417

The *Denver Tribune* reported it, $19,119,007.
The *Rocky Mountain News* (Denver), $25,335,483.
Frank Fosset, it is said, estimates at $18,650,000.

It will be seen by the foregoing that, in 1872, the product of silver began to exceed that of gold in Colorado, and has been largely in excess of it ever since. In fact, in the year 1879, the production of silver is over three times that of gold, so it will be seen the silver interests are foremost at present in the State.

The First Silver Discoveries in Colorado.

Leadville is the phenomenal camp of the year, increasing from a production of three millions one hundred and fifty-two thousand nine hundred and twenty-five dollars, in 1878, to eleven and one-half millions in 1879, which is probably the greatest record in any mining district in the whole world. The production of this camp being chiefly silver, makes the proportion of silver production to that of gold much greater in 1879 than ever before.

The first important silver discoveries were made in the vicinity of Georgetown, in 1864-65, which proved to be rich in silver. This was but the beginning of similar discoveries everywhere. By 1870 there was rich silver districts all over the State, but no considerable production on account of the lack of smelting-works until 1871-72, since which time the production of this metal has steadily increased. In the winter of 1874-75, Congress passed an act enabling Colorado to become a State. It was provided that, upon the adoption of a Constitution by the people of the Territory, that it should be received into the Union as a State, on the 4th of July, 1876.

Colorado Admitted to the Union—The Centennial State.

Accordingly, a Constitution was adopted, and Colorado became the Centennial State. It was provided in the new Constitution that no appropriation of funds should be made for the purpose of State buildings within five years, or until after the first of January, 1881, consequently Colorado has no capitol buildings or State buildings of any sort at present. The first Legislature met in

Denver, in November, 1876, and Jerome B. Chaffee and Henry M. Teller were elected to the United States Senate. From 1876 to the present time, Colorado's history is one of unexampled prosperity. Towns and cities sprung into life and grew with wonderful rapidity. Railroads traverse the State in all directions. It has now a population of two hundred and fifty thousand, with over two hundred towns and cities, and about one hundred millions in taxable wealth. It has over sixty newspapers, of which nearly twenty are dailies; schools, colleges, churches, and some of the best educational institutions in the West. What wonderful achievements in only twenty years! Ten years ago it had not a railroad, now there are fifteen hundred miles of road within the limits of the State. Three great through lines bring traffic and travel from the East. The Union Pacific, Kansas Pacific, and Atchison, Topeka and Santa Fe. Six railroads centre at Denver and extend their arms out to all parts of the State.

Though Colorado is for the most part a mountainous country, she has other rich resources besides those of her gold and silver mines. She possesses a vast area of coal deposits, and great facilities for stock raising. Wherever irrigation is practicable, her soil can be made to produce abundantly. Her stock ranges are of vast extent and not excelled in quality of nutritious grasses. Cattle winter out on her plains without attention.

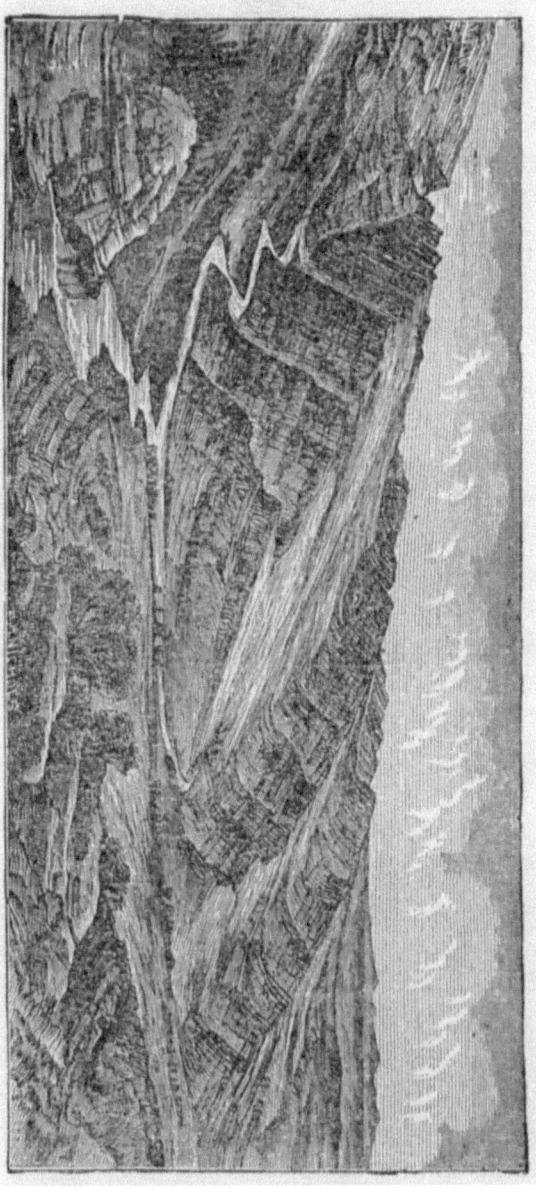

NORTHERN SLOPE OF UINTA MOUNTAINS.

The Uinta Mountains are one of the ridges of the great Rocky and Sierra Nevada Ranges. They sweep through the Colorado and Utah sections, affording great diversity of scenery, and showing the characteristics of the general mountain systems of Colorado.

She has about nine hundred thousand head of cattle and not far from two million five hundred thousand sheep, the combined value of which is not less than eighteen millions of dollars. The business of stock and sheep raising has been a very profitable one, and will continue to be, and will increase to wonderful proportions. There are still thousands of acres unoccupied by ranches or cattle ranges, and the new treaty made by the Secretary of the Interior with the Ute Indians, lately ratified by Congress, will open up for settlement a vast extent of fertile country, exceptionally well adapted to stock raising, west of the Rocky Mountains. The country possesses a milder climate in winter, broader and richer valleys, and more level plains, together with more frequent showers than is found on the Eastern slope. In addition to these resources, the New Gunnison country and the Ute reservation is rich in mineral resources and coal, and the future and prospective wealth of the country, which is soon to be opened up by railroads, can hardly be overestimated.

CHAPTER XIV.

ONWARD TO THE GUNNISON COUNTRY OVER THE BACKBONE OF THE CONTINENT.

RETURNING from our trip to the Black Hills, we take up the narrative of our journey, where we dropped it at Sidney, and pass on over the Union Pacific Railway to Colorado and the Gunnison country, and over the snowy range to the land of promise.

The plain along the railway, west of Sidney, presents the same dry, sterile and barren appearance as farther east; yet it is a great cattle country, and herds are hardly ever out of sight. Droves of antelope, in numbers of from four to eight, are plentiful, and can be seen galloping across the country very frequently, and often a solitary horseman or "cow-boy" can be seen in the distance cantering along on his pony, who looks lonely enough in the vast labyrinth of plain which surrounds him.

In about three hours after leaving Sidney, we came in sight of the grand old Rocky Mountains, covered with snow, and, about this time, passed through one of the longest snow-sheds on the Union Pacific Railway. Our train soon after pulled into Cheyenne, five hundred and sixteen miles from Omaha. Elevation, six thousand and forty-one feet, and having a population of four thousand five hundred. It is well built, partly of brick, and is a busy, bustling town.

From Cheyenne to Denver, via the Denver Pacific Railway (one hundred and six miles), the snow-capped Rockies were ever in sight, presenting to the eye a grand spectacle, in striking contrast to the dry desert plain we had just crossed, and which, in fact, we were still crossing. Mountain rises above mountain, peak above peak from the foothill westward in the distance, their summits white with snow and their peaks fairly touching the clouds, presents a picture which few having seen can ever forget.

The country along the road south of Cheyenne is the same rolling, barren, dried-up pasture-land, which we had seen farther east; no water, no springs, no trees, no grass, but, if anything, more barren and sandy than any section we have ever seen. Yet the country is all taken up with ranches, and frequent herds of cattle are seen.

Long's Peak rises before us a monument of greatness, its summit touching the very clouds; a cool, delightful, health-giving breeze blows from off these snow-capped peaks, which in conjunction with this elevated atmosphere, is exhilarating in the extreme; yet with so much snow in sight, the weather this day was very warm, and we almost felt the need of summer clothing.

As we approached Greeley, in the South Platte

Valley, the country grew better, and we began to see evidences of civilization; an occasional farmhouse dotted the plain, and vast areas of plowed fields met our view. We now frequently crossed great irrigating canals, many miles in length, which bring down from the snowy range the very essence of life and vigor to the growing crops. Indeed, without irrigation, the farmers of Greeley would be unable to raise even a scanty subsistence, except in the way of cattle, which here, as elsewhere in Colorado, is a leading industry. Greeley claims three thousand five hundred population; its elevation is four thousand four hundred feet. It has not a single saloon or bar in the whole city; it is, in fact, a cold water town. The founders of the city prohibited the sale of liquors upon these lands by inserting contracts to this effect in the deeds given for lots, and parties violating this rule forfeit their titles.

It was *very dry* in this section (not on account of above facts), very little rain ever falling. A gentleman told us that it had been about eight months since rain had fallen, though snow had fallen within the time. We arrived in Denver between seven and eight o'clock in the evening, in time to get a glimpse of the city by daylight.

Denver is six hundred and twenty-two miles from Omaha, and about two thousand miles, more or less, from New York. Its elevation is about five thousand and forty-eight feet. It is pleasantly

PEAKS ON THE GREENE RIVER, IN WESTERN WYOMING.

situated on a gentle slope facing the South Platte River, surrounded by a sandy plain, in which there is plenty of alkali and cactus—to all appearance a desert—yet which with irrigation produces abundantly. It is a busy, bustling city, the streets being almost blocked at times with the rush of travel. The city seemed to be full of strangers; the hotels were filled to overflowing; one hotel turned away over a hundred guests one night, being unable to keep them. The hotel arrivals had been nearly six hundred daily for a couple of weeks, and still they came. Many of them were "tender feet" and "pilgrims" from the East, bound for the mining camps of Leadville, Rico, the San Juan and Gunnison regions. Hotels and boarding-houses were thriving, and the prosperity seemed to extend to merchants and tradesmen of all classes.

Wishing to hire a room, we inquired prices where we saw a shingle out, "Rooms to Let," and were told that the rooms had just been taken at twenty-five dollars per month—a small room, with one bed—and the landlady said she had received forty applications for it up to three o'clock P. M.

To all appearance there was a vast army of strangers here, crowding the hotels, thronging the streets, a restless, surging, eager mass of people, anxiously waiting for the snow to melt in the mountain passes, so they could reach the Gunnison country, which appeared to be the objective point of two-thirds of all the people we met, and to which all the world seemed to be going *en masse*. Here they were waiting for something to "turn up," or anxiously looking for employment, which many were seeking and few found, as there were a dozen applicants for every position save in the line of skilled mechanics alone; wages were low and board was high. Mechanics got two to two dollars and a half or three dollars per day, and paid seven dollars per week for board. Other laborers got from twenty to twenty-five dollars per month and board. We saw one poor fellow, with a sad countenance, inquiring "where the employment office was." Another sorry-looking young man came to our hotel, inquiring if any more cooks were wanted, and was answered "none." By the way, there are few women in this city comparatively, and the cooks are all men. The cooking and waiting at the hotels is all done by men; these men get from twenty-five to thirty dollars per month, while women as chambermaids are wanted and advertised for constantly, at the same wages per month. The city is well built, mostly of brick, and some costly buildings were being erected. Lieutenant-Governor H. A. W. Tabor was building an elegant block of stores and a mammoth hotel, to cost probably from three hundred to five hundred thousand dollars; and another hotel, equally as large, was being built by a stock company, to be called the "Windsor."

The Denver Post-office.

Denver needs a new post-office, as the present facilities are inadequate to the vast number who have to be waited on daily. Here we were daily witnesses of sights to us most strange. Each evening, after the Eastern mail was distributed, there was to be seen the greatest crowd of people which it was ever our experience to see at a post-office. There are two delivery-windows, one for letters directed to names from A to L, and the other from M to Z; to each of these windows there was a line of men—single file, and packed closely, frequently from seventy-five to one hundred feet long, if it were in straight line—waiting for their turn to get mail. Frequently we had to wait an hour before we could reach the delivery-window. We were told that at the Leadville post-office even a larger crowd is found at the window.

At the banks one has to wait in a similar manner, and the stores, shops and other places presented the same crowded appearance, and the hotels were *more* than full, and for sleeping arrangements most of them have what they term a "corral," *i. e.*, a large room or hall, with many rows of bunks close together, with a capacity to accommodate a "herd" of people when the rooms are all filled. Even the barber shops were so full as to cause us considerable detention in getting our shave.

Just outside of the city people were living in

tents and shanties of all descriptions. A little calculation gives some idea of the amount of money which was coming here from abroad, and being spent monthly, which helped to keep up the present prosperity. Here were five hundred and fifty arrivals daily; they spend for board from one dollar and fifty cents to four dollars per day, besides this many were spending money freely for drink and cigars, and at the theatre, and some were buying blankets and supplies. It is probable that five dollars per day was less than the *average* expense of each arrival. But, say five hundred and fifty arrivals, at four dollars per day, equals two thousand two hundred dollars per day, equals sixty-six thousand dollars per month left there by strangers alone.

There was a great excitement about the Gunnison country, and even the old gray-headed miners of twenty years ago, who had resided here so long, were all more or less affected by it, many of them were going, and here in Denver we did not hear much else but Gunnison. The papers were full of it, and we heard it at every corner on the streets. "Prairie schooners" passed every day, with cards out, "Passengers for the Gunnison," and Gunnison city lots were offered on "speck" by the real estate dealers by the hundred. There was certainly to be a big stampede to the Indian country this summer. It seemed strange that people would rush to such places in a regular "stampede," repeating the history of "'49" to California and that of ten years later to Pike's Peak. The result of all this will, of necessity, be much hardship and suffering.

The hardships and trials that men will endure for the sake of prospecting for gold, is wonderful. Hundreds of people had already gone to the Gunnison, and were camping out in tents on snow which was in places two feet deep. An old prospector told one of our party that at one time in his experience he lived in the San Juan country *two* months on "ruta baga turnips." He remarked that it "didn't make him very fat, but it was better than nothing."

The streets of Denver are very dusty and dirty, notwithstanding they sprinkle constantly. It being very dry and sandy, the wind brings in a regular shower of dust and sand, which is so full of alkalies that it parches the lips and causes the eyes to smart, and is far from agreeable, yet the clear sky and dry, pleasant weather is delightful, and one soon gets accustomed to sand-storms and will dodge a cloud upon its approach by stepping inside a shop or behind a corner.

The "almond-eyed" race are well represented in this city, one sees them everywhere. Near our hotel, from probably the poorest and oldest shanty in the city, hung a sign, "Sing Lee, Laundry." Wishing to view the premises, and having some collars to wash, we rapped at the door of Sing

Lee, and were greeted by "Come in," spoken in good English; we opened the door, and to our surprise, in a little room, twelve feet square, saw five Chinamen busy as bees, washing, ironing and folding clothes,; one was upon his knees, bending over a low box, on which was spread a shirt which he was rubbing hard with some sort of a brush, dipping it occasionally into water. All about the room, in every conceivable shape, and occupying nearly every foot of space, except that occupied by the stove, were hanging or lying in piles clothes of every description.

Everything in the line of bedding or furniture was of the most squalid and dirty sort. We inquired prices of washing, and found them to be fifteen cents for shirts, five cents for collars and handkerchiefs, and five cents for stockings. Accordingly, we took Sing to our room to get our bundle, when it occurred to us to offer him a bargain. We had with us a couple of boxes of provisions, containing cake, cookies, biscuit, bread, and, in fact, the best of everything needed for white men to eat; this we showed Sing Lee, and offered it to him for washing *two collars*. But, ah, no! Sing had rather eat boiled rice, *stewed rat*, and imported Chinese supplies, than such *trash*, and he shook his head, and answered, "No! no!" However, we must say that Sing Lee did us a good job in washing, and shirts and collars wore the finest possible gloss and finish.

Denver City.

Denver is, undoubtedly, the best built city between St. Louis and the Pacific coast. Brick and stone have been the materials chiefly used in its construction, and very few of the old-time, scrawny or shabby wooden structures remain to be seen. Her growth has probably been more rapid, and her prospects are, perhaps, more brilliant than any other city in the Far West. The far-famed Leadville is the only town which can show a record of growth to compare with Denver, and, of course, the former can never become a commercial centre or distributing point equal to the latter.

Denver is situated at the junction of Cherry Creek with the Platte River, and was once—strange as it may seem in a country ordinarily so dry—visited by a terrible flood from the former stream. A terrific thunder-shower in the night raised Cherry Creek to a huge mountain torrent, washing away a portion of the city, and causing the loss of many lives and a sacrifice of thousands of dollars worth of property. At the time of our visit, Cherry Creek was a dry water-course, without a drop of water. The pioneer gold-hunters found gold in this stream in the vicinity of Denver, and also in the bars of the Platte River, which, undoubtedly, caused the town to be laid out on the present site. Even yet it is said gold can be found in the gravel taken from the cellars in some of the lower streets in the city; although, of course, in very small quantities.

The floating population of Denver is always large, and it is claimed that no other city of double its population does half the hotel business of Denver. In almost an equal degree is the mercantile and jobbing trade remarkable. Single firms of wholesale grocers sold one and a half millions' worth of goods last season, and jobbers in other branches of business nearly as much. Denver is the receiving and distributing point for an immense industry and population, while half a dozen railways centre there and help carry on the work.

During the year 1879 not less than one hundred and sixty thousand tons of freight were received at Denver, and over one hundred and twenty-five thousand tons were forwarded. The sales of merchandise reached seventeen millions of dollars. The value of Denver's real estate and personal property is put at twenty millons of dollars. The receipts at the Denver post-office for the past year were about fifty-five thousand dollars, mostly for stamps, and there were money-orders issued to the amount of about two hundred thousand dollars.

The hotel arrivals in Denver, for 1879, was not less than seventy-five thousand. The population of the city, according to the census of 1880 (at the present writing not fully completed), is not far from thirty-four thousand. Denver organized a Board of Trade in 1867. The Union Pacific reached Cheyenne the same fall, bringing the new city within reach of railway communication.

During the summer of 1870 the Denver Pacific was completed from Cheyenne to Denver, connecting the metropolis of the plains with the outside world. In August, 1870, the Kansas Pacific reached Denver, making two through lines to the East. Other railways followed in rapid succession, extending their branches to the mining camps in the mountains, and Denver's "boom" began in earnest, which has never ceased up to the present time, and she began a career of unexampled prosperity.

On to the Gunnison—Over the Snowy Range.

After stopping in Denver for two weeks, waiting for the snow to melt in the mountain passes, before venturing to cross the Rockies, we took the Denver and South Park train for Buena Vista. The road for the first twenty miles from Denver is over a level country, along the base of the Rocky Mountains, near the foot-hills, and is through a country of green fields and irrigating canals, rich farms and gardens, interspersed with desert sands and alkalies. Suddenly our train rounds a curve and plunges into Platte Cañon, up into a very labyrinth of hills, and peaks and jagged rocks; up a rapid stream (the South Platte River), which has cut a crooked, winding channel out through the mountains to the plain below. At the mouth of this cañon (pronounced *canyan*) an English company, with a large capital, have begun an immense ditch, and are blasting out a watercourse along the precipitous, rocky sides of the cañon, purposing to take water from the river several miles up, and carry it a hundred miles out on the plain, where they have purchased large tracts of land, which they wish to irrigate. Let our Eastern friends recall to their minds the picture of the road along the Lehigh River, from White Haven south, near Mauch Chunk, Pa. Only please imagine the stream to be smaller and much more crooked; the hills to rise from five hundred to a thousand feet higher, with less timber upon them and much more precipitous, and instead of a few miles, makes it at least forty miles in length, and with this picture in your imagination, let us try to describe to you Platte Cañon. Here is a gulf worn into the rocks by the Platte River in the ages past, narrowing up in places to forty or fifty feet, and so crooked and winding that in ascending it we traveled toward every point of the compass, and could see the engine ahead of us from one side or the other almost constantly from our car-windows. Much of the way the sides are so steep, and rugged and barren of earth as to offer no foothold for the telegraph poles, for which have been substituted iron bars, inserted in holes drilled into the rocky sides of the cañon, to which the wires are attached.

These walls are mountains, of unequal height, varying from five hundred to over a thousand feet,

WINNIE'S GROTTO—A SIDE CANON—Walls 2,000 Feet High

and from the vertical, in places, to a gradual slope, and whose cracked, jagged, and fissured sides and overhanging rocks look down on the traveler from a dizzy height above, with seemingly little to hold them in place, or keep them from falling into the abyss beneath. To Eastern people it is a grand sight; to Coloradoans it is but tame compared to other and grander cañons in the State. The railway is a narrow gauge, and is therefore adapted to the sharp curves and winding, narrow path it has to follow.

We passed, on our way up, many miners with their packs and blankets, footing it to the mines; bound for Leadville and the Gunnison.

At Deane's Station, a few miles up the cañon, our train was detained by an accident to a passenger train ahead of us, which was on its way down. The engine had encountered a huge pine-knot, which had rolled down upon the track, throwing the engine down the banks and crushing the poor fireman beneath his engine and slightly injuring the engineer, who saved himself by jumping. Strange to say, but one passenger truck left the track, and no passengers were hurt. While waiting at Deane's Station for the road to get cleared, we bethought us of our fish-lines, and learning that the Platte was filled with trout, we cast our hooks into the stream, and one of our party soon caught a speckled beauty, weighing half a pound. We were, however, less fortunate, and were obliged to quit the pastime without being successful.

We soon passed the scene of the accident, where we saw the wrecked engine in the creek bed, with the poor fireman still fast beneath, his hand still on the lever to the air-brakes, where he had died bravely at his post.

It was a day of accidents and delay; a few miles further on we were detained by a broken bridge, and a delay of two hours ensued. Perhaps there may have been "luck in leisure," for while waiting here the passengers left the train, and gathering in groups were soon heard conversing on that all-important subject—*mining*. This one was superintendent of the "Bald Eagle" mine; another was exhibiting a certificate of assays, made from some of "the richest claims" he had ever seen. Another was a capitalist, no doubt going to invest in some of the Leadville bonanzas,—who wouldn't, he said, "give a cent for an assay," but must have *"mill runs"* to test the ore. A story was related, to prove his point, of the successful Chicago assayer, who always gave high assays, and of course very much pleased his patrons, who, to test him, sent in a piece of grindstone, and to their surprise found it *run a hundred ounces of silver to the ton!* The mining superintendent then related the fact of a Leadville assayer finding *fifty ounces* to the ton in an old jug handle; and of another Leadville expert, who, when given a specimen of ore, asked, "Shall I make it run rich in silver or in gold?"

We soon passed the bridge safely, and were climbing up toward Webster summit at a grade of two hundred and thirteen feet to the mile, which, when it is reached, is one of the highest points a railroad has yet climbed in the United States, or ten thousand one hundred and five feet (10,105). Here we wound up the mountain side on one side of the cañon, then crossed by a sharp curve to the other side, retracing our way, yet slowly creeping up the mountain to a higher level, where we could look down from a dizzy height into the valley below, and plainly see the track by which we came on the other side, two or three hundred feet below us: up, up we rise by a winding, crooked path, so narrow as to cause fear of sliding off our miniature track, till it fairly makes us dizzy to look below. Suddenly we leave the cañon and strike across a sort of table-land, yet still climbing a heavy grade, and snow-capped mountains still rising high above us. Finally the summit is reached, called "Keneshoa Divide," and we begin our descent down toward the waters of the Upper Arkansas.

Soon a grand sight met our gaze as we emerged from among the mountains out into plain sight of South Park. There is a rolling, partially level valley, surrounded by snow-clad mountains on every side, and is many miles in extent. We crossed this park, thence down into the valley of the Arkansas, and in due time arrived in Buena Vista without accident.

Buena Vista—The Road to Leadville.

Buena Vista is a new town of mushroom growth, situated near the head-waters of the Arkansas River, and lying on both sides of Cottonwood Creek, a tributary of the same. Its altitude is eight thousand feet, and distance from Denver about one hundred and thirty-five miles. This town five months ago had only three buildings, and it now contains three or four hundred cheap wooden structures, some well-built hotels, saloons, and gambling-places without number. It claims a population of one thousand five hundred. The town is in the midst of a park country, surrounded on all sides by magnificent snow-clad mountains. A little south-west Mount Princeton rises up grandly to the height of fourteen thousand one hundred and ninety-six feet, and other peaks raise their whitened summits almost as high.

Tents greet the eye along the streets in every direction. Large canvas-roofed buildings and tents, filled with merchandise,,and crammed full of goods. Mammoth tents, transformed into freight-houses, and full of freight and supplies *en route* to Leadville and the Gunnison country, are everywhere seen. At the depot we saw seven four-horse stage-coaches loading with passengers for

Leadville, besides other vehicles taking in passengers for Alpine, Pitkin and Gunnison City. Eight, and sometimes nine persons are crammed inside these Concord coaches—which look as if four persons would sufficiently load them inside—while four or five persons are put on top of the box. The fare to Leadville from Buena Vista is five dollars, distance thirty-four miles. Six and eight-mule freight-teams, hauling two canvas-covered wagons, one hitched closely behind the other, laden with freight of all kinds for Leadville; jacks and pack-mules, loaded with their packs for the Gunnison, were constantly moving out. At the freight depot tons upon tons of base bullion (silver and lead), in bars (or pigs) of from fifty to one hundred pounds each, were closely piled upon the platform for fifty feet in length, which was fairly breaking down with the great weight it bore. These bear the stamp of prominent smelting works, and are being shipped East, where the silver will be separated from the lead. Hundreds of sacks of high grade carbonate ore, of perhaps a hundred pounds each, lie here on the platform awaiting transportation also.

It is astonishing what a vast number of people are still rushing to Leadville; yet we were told by reliable men from there that there were perhaps ten thousand idle men in that city, many of them without money or means of subsistence, depending upon "luck" for a square meal, and on their blankets for lodging. We saw men *en route* for that place from every State from Maine to California. We encountered many from the Eastern States, as well as from California and Nevada; even Idaho Territory is represented in the list. People in the East little know of the tricks of towns and railroads, by which they get up a "boom" and an excitement, as may suit their purpose. Leadville has reaped a rich harvest from one of those "booms," yet we are informed by parties who should know, that the merchants and business men of that place once raised by subscription not less than five thousand dollars, which was used in advertising that camp far and near throughout the country. It seems almost incredible, yet it is no doubt true. But Leadville has had its "boom," and is doomed. Rents have fallen off very much, and real estate cannot be sold at its former value. Yet they are still striking new mines, and the daily reports that some poor prospector has struck "it rich," will still attract fools enough to keep up the "boom" a while longer. Buena Vista has some mines in the surrounding mountains, though none sufficiently developed to create any excitement as yet. About seven miles south-west a new mining camp, called Cottonwood, is getting up some excitement, but as yet little is known in regard to it. Alpine, eighteen miles south, is quite a lively mining camp, and has some valuable mines. Buena Vista has been a lively town for the past four months, being the terminus of the Denver and South Park Railroad, and the nearest station to Leadville. But the Denver and Rio Grande Railway will have completed a road up the Arkansas, from Cañon City, past this place to Leadville, by July 1st, when this town will have to decline. The buildings here, as before stated, are of the most temporary sort, except a few of them. The hotel where we stopped was plastered with building paper, tacked to the walls, and ceiled and whitewashed overhead with thin white muslin, fastened in the same manner. This was the case with all the rooms, except those where the rough boards were uncovered. We got good accommodations, however, and were well fed for two dollars per day. Some of the hotels, though, ask three dollars.

We started on a bright, pleasant morning from Buena Vista across a rolling park or plain, toward Mount Princeton, around whose lofty snow-capped summit hung some dark, stormy-looking clouds, partially hiding it from view. Occasionally its glistening summit could be seen above the clouds, producing a picture grand beyond description.

We had expected that the mountain was but a couple of miles away, and after having gone that far without approaching any nearer to its base, we asked the driver how far it was from us still, and learned that it was six or seven miles. We found later that we could place but little dependence upon our eyes to measure distance in this light atmosphere. At about nine miles from Buena Vista we passed the base of Mount Princeton and came to a sharp precipice, where we could look down into the cañon of Chalk Creek, into which we descended by a narrow and precipitous road, so steep that we much preferred walking. To our right rose some white, rugged, chalky-looking mountains, a thousand feet above us, called Chalk Cliffs. To our left Chalk Creek came dashing and foaming down the cañon, along which a gang of men were at this point constructing an irrigating ditch of great length, to carry water down to the plain below. The Denver and South Park Railroad Company were also grading up the cañon for their extension up to Alpine and through the tunnel to Pitkin and Gunnison cities.

Chalk-Creek Cañon is narrow, and the mountains that rise up abruptly on either side are a white, decomposed granite of a chalky color, fissured and seamed in every direction, with many curves and holes worn into the soft perpendicular rocks by the elements, giving the cañon a wild and picturesque appearance. Near the mouth of the cañon there bursts out from beneath the Chalk Cliffs the Hortense Hot Springs, the waters of which are warm enough to cook an egg, and are said to be possessed of valuable mineral qualities. There is also a bath-house near by. The road up the cañon as far as Alpine is quite good, though

MOUNT OF THE HOLY CROSS, COL.

it was terribly dusty. Our fare from Buena Vista to Forest City, twenty-three miles, was three dollars each, and our conveyance a lumber wagon with a spring seat.

We reached Alpine about noon; found good hotel accommodations for seventy-five cents per meal, or two dollars and twenty-five cents per day. Alpine is properly named; it can properly be called the Switzerland of America. Surrounded by high mountains on every side, which rise from one to two thousand feet above the place, their peaks reaching far above timber line, it is certainly the most genuine Alpine scenery in America. The town is building up rapidly, and contains, perhaps, a hundred or more houses and log cabins, and one or two good hotels. Its elevation is about nine thousand two hundred feet. There are many prospect holes and several good mines in the vicinity of Alpine; the Tilden mine, of this vicinity, selling a year ago to Eastern capitalists for a very large sum. The road from Alpine to Forest City, which is four or five miles further up the gulch, is not good, but very rough. We frequently got out and walked up the rocky hills, but we began to experience some difficulty in breathing in this light atmosphere. An oppressive feeling about the lungs, and a sort of giddy sensation about the head, were some of the symptoms we experienced. Few Eastern people realize the great change in the density of the atmosphere between the average elevation in Pennsylvania and most of the Eastern States, and the elevation here among the Rockies of ten thousand feet and upwards. People should get acclimated to such high elevations gradually, and not rush into the mountains to a great elevation too suddenly from a lower atmosphere, as the sudden change often produces disease, and frequently pneumonia, in connection with the necessary exposure to a colder climate.

We reached Forest City about four o'clock, and put up at the best hotel the place could afford, which was, in fact, the only one, at three dollars per day. The hotel was a rude log house, recently constructed, and but half finished. Outside the cracks in the logs were chinked and plastered, but inside only the bare logs, without ceiling or plaster of any sort. When we asked about a bed and a room by ourselves, the landlord told us the best he could do was to give us a bed and draw a chalk mark around us for a room. We accepted the situation.

We were complaining of shortness of breath, owing to the high altitude, and an old doctor, boarding there, volunteered us some medicine. He handed us a vial, saying, "Take a teaspoonful." We took a swallow, and the old doctor, watching the bottle, asked suddenly, "What! you didn't swallow any of that, did you?" We confess it startled us, and visions of poison by mistake, antidotes and stomach-pumps flashed through our mind like lightning, as we hurriedly inquired what we had taken, and was answered, spirits turpentine. Of course we felt relieved. It was diluted with water, and therefore harmless.

Forest City is properly named. It is a town of log cabins, tents and shanties in the woods, among the stumps and rocks, and among the trees, which are just beginning to be cut away; with bold and precipitous mountains on either side, partially covered with pine and spruce. From Forest City a trail leads up one of the gulches to the railroad tunnel, and thence across the range to Pitkin. We had been told that it was only twelve miles from Forest City to Pitkin; we now learned that it was full eighteen, having increased fifty per cent. since leaving Buena Vista, and this distance must be walked. We were in a dilemma. We had heavy satchels and a roll of blankets, and to carry them eighteen miles seemed impossible. But hearing that a woman had walked over the range in a snow-storm, we took courage and determined to make the attempt.

The greatest excitement seemed to prevail in all this section in regard to the great Gunnison country, and hundreds of miners and prospectors, on foot, and loaded down with their packs, containing blankets, tents and cooking utensils, were constantly passing, bound for Virginia City, Pitkin and Gunnison, over the range. Some were driving ahead of them jacks packed with supplies.

Some were also returning, bringing rather discouraging reports from the country. Some poor fellows were tired out and sick, and compelled to lay by at our log hotel from this cause Men with their pantaloons saturated with mud and water up to their hips, returning from Pitkin, were constantly met with, showing evidence of the horrible condition of the road to the summit. After considering the matter carefully, in connection with the condition of the roads, we resolved to stop one day in Forest City, and not attempt the passage till the following day, when we could secure a couple of saddle-horses, accustomed to the trail, to help us seven miles on our journey—as far as the tunnel.

During the day we had our first excitement in prospecting for gold. We and our friend climbed the steep mountain side near by, determined to "strike it rich." We very soon encountered several prospect holes, where tunnels were being driven into the hill in search of mineral, and we were encouraged. Here was mineral, sure. Accidentally picking up a stone in our path, we examined it, and to our satisfaction it contained specks of gold. Our friend examined it with his magnifier, and became excited. "Yes, yes! that's gold, you bet! Where did it come from? Where was the lode?" were his eager inquiries. We searched diligently, but in vain; we discovered no vein of ore and no deposit, save a large boulder of the same kind of rock, from which our specimen had evidently been broken. Delighted with our find, we took our specimen to town, and learned from an old miner, to our great chagrin, that it was simply iron pyrites, and worthless.

There are several mines and prospect holes near Forest City, but much of the ore is chiefly Galena and iron pyrites, not carrying much silver. However, there are said to be some rich mines four or five miles up Grizzly Gulch, which are profitably worked. A miner took us into his tunnel seventy feet, in which the face of his drift fairly glistened with pyrites. Of course he wished to sell his mine, and only asked the moderate sum of fifteen thousand dollars. However, he has a good streak of lead ore, which he claims assays well in silver, and he may yet strike a genuine bonanza.

A half mile above Forest City, Chalk Creek Cañon divides into two gulches or ravines, down each of which flows a rapid stream. A trail leads up the right hand, or eastern gulch, over Alpine Pass to Virginia City and Hillerton, two good mining camps in Gunnison County, and near which is the celebrated Gold Cup mine, and also the Tin Cup and Silver Cup districts. During the snow blockade, the past winter, provisions became very scarce and high in those camps, and we were informed by a man who went up there very early in the spring from Forest City, with potatoes and flour, that he got fourteen and a half cents per pound for flour, and sixteen and a half cents for potatoes, and as the journey had to be made with snow-shoes and a hand-sled, very little could be sent in even at those figures.

Walking Across the Continental Divide.

Bright and early on a Saturday morning we mounted our saddle-horses at Forest City, for which we paid two dollars each for a few hours' ride, and struck into the trail that led up the left-hand gulch toward the summit where the South Park Railroad Company is to pierce the Continental Divide. The air was crisp and frosty, and the sun was as bright as ever a May morning could find it. We followed up a very rapid stream, and gradually rose higher and higher, and the mountains became less lofty as we ascended, until they were mere hills compared to those we had left behind. We soon encountered patches of ice frozen during the night, and everywhere drifts of snow, and a little farther on the ground became covered, and we found ourselves riding over a trail of beaten and packed snow of from two to five feet deep. Occasionally our horses would break through with one foot and recover without falling, but finding no bottom. Many places in the deep ravine it may have been ten feet deep, yet the road was so hard that we passed over it in safety. A journey of seven miles over marshes, fording streams, over rocks and fallen trees, and over the frozen snow, brought us to the tunnel. When in sight of it we could see men beyond climbing the trail that leads over the range. In the distance they looked like little dwarfs, with their packs on their backs, creeping slowly up to the top. At the tunnel they have a rude camp of log cabins, nearly buried up in the snow, in which the workmen live. The cabins, which have been built this spring, apparently had been erected where seven or eight feet of snow had been removed, and there remained still snow as high as the roofs. The eastern end of the tunnel is in about one hundred feet, and the western end has just fairly begun. It will be, when completed, something over one thousand six hundred feet in length. Dismounting from our horses, and tying up the reins according to order, we started them back alone on their journey. We then began to ascend the slippery trail, over snow apparently ten feet deep in places. But the path was so steep and the air so thin, that we could only go a few steps without sitting down to get breath. The sun shining brightly upon the snow, troubled us somewhat with snow-blind. Many of those passing over we noticed wore glasses to avoid this. Slowly we worked our way up. The distance was not long from the tunnel, perhaps less than half a mile, and at nine o'clock we had reached the summit of the grand old Rockies, over twelve thousand feet above sea level. What shall we say to do justice to our view from the summit? Notwithstanding our great height, around and above us rose magnificent mountains still a thousand feet higher; and beneath us, to the east and to the west, in the distance we could see deep cañons, so deep as to almost make us giddy to look into them. To the east the stream we came up pours its waters into the Arkansas, and thence into the Gulf of Mexico. To the west Quartz Creek sends its waters into the Gunnison River, thence into the Colorado, and finally into the Pacific.

We soon crossed the Continental Divide. It is a narrow, nearly level flat upon the mountain, perhaps twenty rods in width; and we began to descend the precipitous mountain-side as abruptly as we had came up. We soon reached the mouth of the tunnel on the Pacific slope, passing it in the gulch beneath us, and struck into the trail in a bank of snow, where we experienced the need of snow-shoes. Frequently we would sink into snow to our waists, and would recover to sink as suddenly again, sometimes with one, and often with both limbs. At one point our friend who accompanied us took a plunge into the chilly depths up to his very neck, and found much difficulty in extricating himself from the painful situation.

It was a toilsome journey for a couple of miles, but luckily the snow then disappeared, and we had no further trouble from this source. We were soon caught in a snow-squall, however, during which it snowed briskly for a few minutes, but was of short duration, and the sun was shining again as brightly as ever. We came to a point after a

Lost.

time where we lost the trail in the snow, and were fairly lost. After hunting awhile we saw below us some men toiling up the hill, and going toward them found the trail. Our path lay from thence down into a deep cañon, where the trail was so steep as to almost prevent a horse from climbing it, yet which bore evidence of horses and pack animals having passed along. Over rocks and boulders, fallen trees and brooks through a forest, it led us until we struck a beautiful stream in the valley, called Quartz Creek. We overtook some men, when part way down, who were carrying heavy loads of flour and tent cloth on their way into the valley. They had brought a horse over the range, and getting stuck in the snow, were compelled to stop and unload the horse, and lead him down the difficult trail, where they left him and returned, carrying down the pack themselves.

Beaver Dams.

Following down Quartz Creek, which occupies a lovely little valley, in which the flats are covered with grass and the hillsides with pine timber, we frequently saw evidences of beaver dams along the stream. We found many trees gnawed down by these intelligent creatures, and many remains of dams constructed by them. We saw numberless trees, from five to seven inches through, which were gnawed partially down, and others which were clear down. Some of these were some little distance from the stream, and were a heavy load for two or three men to carry. How the beavers succeed in getting them into the stream at that distance is more than we can imagine.

Arrival in Pitkin.

We arrived at Miller's ranch, where there are two or three log cabins, about noon, which were about the first evidences of civilization, and where travelers are lodged and entertained when wishing to stop awhile on their journey. There are two or three mines in this vicinity. The "Margaret," owned by Miller & Hall, was discovered about a year ago. It seems Miller came along beneath the ledge and camped for the night. Finding float near by next morning, he traced it up the hillside a few rods, where he started a tunnel, striking a vein of silver lead (Galena) ore, of considerable value. The parties claim to have been offered twenty thousand dollars for their find. Quartz Creek is a very rapid stream, with a succession of falls and rapids, and we descended from our high elevation very fast to a lower one. When down in the vicinity of Pitkin the mountains rise less abruptly, and not so high, and become heavily timbered with a species of yellow pine. The snow had all disappeared, except in a very few places, and the weather was warm and delightful. Little brooks of clear, pure water came pouring down from the hillsides occasionally, from which we drank with delight.

The first evidences of our near approach to Pitkin was the sight of a steam saw-mill, and some log huts in the gulch above the town. The valley here widens out to one-fourth of a mile or more, and Pitkin is situated on a level, grassy flat, a mile or more in length. It is a lovely situation for a town, being on a beautiful grassy park, in which there is no timber, yet the surrounding hills are thickly covered with pine and spruce to their very tops. Quartz Creek passes one side close to the mountain, and an irrigating ditch carries a stream of clear water the whole length of the town, down along the side of Main Street. Main Street is thickly studded with buildings in all stages of erection. Tents and log huts are everywhere to be seen, and for a town of only four or five weeks growth, shows a remarkable energy in the great number of buildings which have been erected. About five weeks ago three or four log cabins were about the sum total of buildings in the place. Now corner lots sell for eight hundred dollars. Its elevation is about nine thousand five hundred feet.

CHAPTER XV.

PITKIN.

DURING the summer of 1879, a party of prospectors crossed the range and entered the valley of Quartz Creek, and following it down to the beautiful park in which the town of Pitkin is built, went into camp, and began to actively prospect the surrounding hills for minerals. Their efforts were rewarded, and rich "float" was found on the surface nearly in all directions for miles around, indicating the presence of rich ore bodies in the vicinity. A few locations were made during the season, and some assessment work done, and late in the fall the town of Pitkin was laid out and christened with the name of the popular governor of the State.

A few log cabins were built, and roofed with poles and earth, and probably a score of persons wintered in the embryo city. The miners who went out in the fall with their samples of ore were not idle during the winter, and by early spring the news they had carried to Denver, Leadville and the States, in regard to the camp, and the tales of flattering prospects and rich strikes which they related, had created a wide-spread interest in the camp, and hosts of men were ready and waiting, in Denver and Leadville, and other places, to rush into the supposed El Dorado as soon as the weather and snow would permit. The winter proved to be a severe one, and the snow-fall was unusually deep and hung on late in the spring; but, notwithstanding these difficulties, March found many coming into the section on snow-shoes, over six or seven feet of snow in places, and men were found shoveling away the snow and erecting log cabins. By the first of April teams were bringing in goods from Alamosa, via Saguache and Chocketope Pass, over a fearful road, wading through two or three feet of snow in coming up Quartz Creek, and in the streets of Pitkin beating a track through drifts in places much deeper. From this time on Pitkin began to "boom." Neither cold or bad roads or snow seemed to prevent people from coming to this new-born wonder-land in the mountains. We arrived in the place about the middle of May. The snow had disappeared, except in sheltered spots in the woods. The town had been growing but six weeks, yet it contained a vast number of log and frame houses, hotels, stores, saloons and tents. Real estate was high; town lots were selling at seemingly fabulous prices for so young a town. Lots suitable for business purposes, on Main Street, were selling at from four hundred to eight hundred dollars, that during the winter could have been bought for a trifle.

Speculation was rife, and a few were making money by the "rise."

One old gentleman, who came into Pitkin last fall, without any means save his team and less than one hundred dollars in cash, purchased nine or ten lots, paying from five to ten dollars per lot, and holding them until this spring, disposed of them readily at prices ranging from three hundred to seven hundred and fifty dollars, thereby clearing about five thousand dollars. Good hotels not being plenty at the time of our arrival, we took quarters in one of the numerous lodging-houses which appeared to be the fashionable sleeping-places of the town. The house in question was a large tent, eighteen by fifty feet; the floor was the ground, which was, however, not level, and was carpeted with three or four inches of sawdust. A canvas partition divided the sleeping apartment from the rest of the room There was a stove and numerous boxes, trunks, etc., as substitutes for chairs, which constituted the furniture of the room. The bunks on which we slept were made of rough boards, arranged in a row at the sides of the room, with two tiers, one above the other, steamboat fashion. The beds consisted of loose hay, thrown upon the boards, and covered with gray blankets; and blankets and comfits were used for covers. No sheets or pillows were to be seen; coats were universally used for pillows. Price of lodging, fifty cents.

We had walked over the range, and were fatigued, and early in the evening rolled ourselves up in the blankets of the upper tier, and went to sleep. We slept well after our tiresome walk, and in the morning felt refreshed. We went to the "Bon Ton" restaurant for breakfast, where we found two ladies in charge. It was a tent, sixteen by twenty-four feet, with a sawdust floor, like our bunk-house, and the tables reclining at an angle of about fifteen degrees from level. However, everything bore the appearance of cleanliness and neatness, and we had an excellent breakfast of beefsteak, bacon, fried eggs, fried potatoes, corn bread, warm biscuit, butter and coffee, and, in fact, everything essential to a good appetite, and well cooked, all for the sum of half a dollar. We were surprised at such good fare, in so new a town, for so small a sum; but found it to be the ruling prices of the place.

The following day was Sunday, yet the stores and saloons were all open; the sound of the saw and hammer were ringing all day long on the new buildings being erected, and the reports from shots of giant powder, while blasting in the mines, were frequent all day; and the din and rush of travel and freighting through the streets went on as usual. These new mining towns have very little regard for the Sabbath.

Buildings were going up very rapidly; many of them very substantial wooden frame structures.

Three saw-mills were turning out lumber very fast, and were disposing of it as fast as it could be manufactured, and had constantly orders ahead. There were within six weeks erected fifteen hotels, restaurants and lodging-houses, and some forty or fifty business houses and saloons. By the 5th of June, when a count was taken, there were one thousand and fifty people within the city limits, and undoubtedly as many more were camping and prospecting within a few miles around the city. There were one hundred and eighty-six dwellings, four hotels, eight restaurants, twelve saloons, fifty stores, business houses and bakeries; eighty vacant and unfinished buildings, three meat markets, several real estate offices, one bank and one jail. There were in all, including tents and log cabins, about four hundred houses. But with all this population, there were but fifty ladies in the town, and about fifty-five children. A fine school building was being erected; and all this had been accomplished within the short space of two months.

The first sermon by a Christian minister was delivered on June 6th, and was well attended by an audience of miners, who were well dressed, orderly and attentive; and two or three women were in attendance. A collection was taken, and the sum of eighteen dollars given the clergyman for his services.

The city of Pitkin at this time was a fair sample of a new and excited mining camp. Prospectors were flocking in at the rate of seventy-five per day. Probably five hundred locations of mining claims had been staked in the surrounding hills, within a radius of five or six miles. Some very flattering prospects had been found, and very high assays of the rock had been given. Business of all kinds was lively and booming. Some very large stocks of goods had been brought in, and several merchants were carrying stocks of from three to ten thousand dollars. One firm was doing a business of about seven hundred dollars per day in general merchandise, and a hardware firm had sold over three thousand dollars' worth of goods from their pile of freight lying in front of

Prices of Living in Pitkin.

their unfinished store, before moving into it. Hay was selling at five cents per pound, or equal to one hundred dollars per ton; flour, eight cents per pound; hams, eighteen cents; eggs, forty cents per dozen; lumber, forty dollars per thousand, which afterward fell to thirty-five dollars; potatoes, nine cents per pound; butter, fifty cents; rice, twenty cents; dried apples, twenty cents per pound; corn-meal, seven and a half cents per pound; beefsteak, twenty cents; nails, fifteen cents; and coal oil, one dollar and twenty-five cents per gallon. At this time everything had to be freighted from Alamosa, or from points equally as far away, and the freights were two and three-quarter cents per pound in addition to the railway charges of one cent to one and a quarter cents per pound. The time on goods from Alamosa was never less than ten days, and frequently longer, and the time from Denver two weeks.

We soon became acquainted with many miners and prospectors, and found them to be an intelligent, social and jolly class. Though often roughly dressed, and carrying a belt containing a knife and large six-shooter, yet possessing large hearts and kindly dispositions toward those who treated them well. Although the Colorado miner boards himself in a log hut or tent, and perhaps fries his bacon in a dirty spider, yet, as a class, they live well, and purchase the best of food and canned goods and fruits of every kind, without stint. No matter how expensive these articles may be, they are a part of their daily living.

Prospecting.

It was a surprise to us what hosts of miners were here who had been former residents of Leadville, the great carbonate camp. Nearly every acquaintance we made was a former Leadvillian, and Leadville capital was well represented in the business houses, stores and saloons of Pitkin. Verily our prediction was coming true, that Leadville was destined to be diminished by thousands when the rush began to the Gunnison country.

We started early on a Monday morning, a party of three of us, to prospect for gold. To us it was new business; we were, in the language of miners, "Tender-feet," though one of our party had prospected some before. With high hopes and great expectations of striking a Bonanza, we broke for the Hills in a northerly direction, and after walking up a deep ravine for a couple of miles, commenced climbing the mountain. Up, up we climbed, where the ground was so steep in places that we caught hold of trees to assist us in our ascent. We soon began to find "float" quartz, and occasionally some showing signs of ore, with iron and copper stains on the quartz, when the business of prospecting became interesting. We continued our ascent until we had reached an elevation of perhaps a thousand feet above the town, and could look down into Pitkin two miles away.

At the summit of this mountain we found quartz croppings, and excitedly we all began to dig for the vein, while visions of untold wealth flitted through our minds as we began to pry out with our picks particles of ore. We worked with a will. The interest excited in digging for gold will cause men to work with remarkable energy. But we soon found, however, that nothing short of giant powder could sink far into the solid rock which we encountered. Therefore, we abandoned work for the day, and proceeded to erect our discovery stake and locate our claim in due form, and erected a stake, writing upon it as follows:

"THE TENDERFOOT LODE.
"*Located May 24th, 1880.*

"We, the undersigned, of the town of Pitkin, county of Gunnison, and State of Colorado, claim, by right of discovery, one thousand five hundred feet along the course of said lode, vein or deposit, with all its dips, angles and variations, as allowed by law; together with one hundred and fifty feet on each side of the middle of said vein at the surface; one thousand two hundred feet running northerly from this discovery stake, and three hundred feet running southerly from this stake.

   Signed,   "JOHN DOE,
                              "RICHARD ROE,
                              "BILL PLUMP."

We then continued our journey up the mountain, prospecting as we went, until weary of going, we retraced our steps to Pitkin, where we produced our specimens with considerable pride for inspection, when, to our disgust, the miners all pronounced it very poor ore.

Although the weather was very pleasant during our stay in Pitkin, yet the nights were cold, freezing ice frequently half an inch thick in May, and as late as the last of June there were heavy frosts nearly every night. We found that we needed to sleep under many blankets to be comfortable. Pitkin, however, is two or three weeks ahead of other camps in Gunnison County, as the snow had all disappeared, and the roads were in good condition long before the trails were open to Ruby and Gothic cities, except for men on foot.

Pitkin may be considered a healthy place in which to live. There seemed to be a large supply of physicians, but very little for them to do. There had been a few cases of mountain fever, rheumatism and pneumonia, but not more than four or five deaths from all causes up to the time we left.

Mail Facilities.

Pitkin has a daily mail from Alpine and the East, carried on horseback, the time being twenty-four hours from Denver. There is also a daily stage, carrying mails and passengers to and from Gunnison City, the time being about five or six hours

Gambling.

Miners, as a rule, are universal gamblers. There seems to be something about their wild life that especially gives them a disposition to gamble. The fact is, that prospecting is such a game of luck or chance, that they come to regard it almost like a throw of the dice. A prospector sinks a shaft or runs a tunnel, with very little prospect of ore at the surface. He "goes it blind," to use a miner's expression, taking his chances of striking a lead. Hundreds of prospect holes have been sunk in this manner in and around Pitkin. The gambling-houses of Pitkin, like those of other camps, are well patronized, and every evening sees their tables well occupied by men who regularly spend their money in this manner. Many of these are constant losers. One Leadville prospector, who bunked in our tent, lost sixty dollars in a couple of nights, and although a poor man, with a family, he was afterward just as ready to try his hand again.

Wages

were good for mechanics and miners at Pitkin, and laborers were always in demand. Clerks, barkeepers and professional men, however, were not very much in demand, and their pay not high. Carpenters were getting three dollars and a half to four dollars per day; other laborers, two and a half to three dollars per day.

Law and Order.

Pitkin was remarkable as being an orderly mining town during our stay in the vicinity, there being very little theft, or violence, or resort to firearms to settle difficulties of any sort. Person and property was apparently as safe in Pitkin as in an Eastern city, although everybody carried arms, and it was no strange thing to see belts containing revolvers, knives and cartridges carried openly and exposed to public view.

An Incident.

One evening, in the outskirts of Pitkin, on the road to Gunnison City, a horseman was seen riding slowly into the town, driving ahead of him two burros, with heavy packs of mining tools, a tent and camp utensils, the poor donkeys being so tired that apparently they were ready to lie down with their loads at any moment, and the man was with difficulty urging them along. The man was a dark-complexioned, hard-featured character, and his face strongly marked with smallpox. In the distance, half a mile away, there was seen a cloud of dust, and four horsemen were seen approaching, their steeds galloping at the height of their speed. They approached the traveler with the burros, who appeared not to observe them, and separating as they drew near, they surrounded him in an instant, and the cry came clear and distinct, "Hold up your hands," as four trusty revolvers were leveled at his head. The man could do no more than to obey, as with a scared look he stammered, "Why, what's the matter? I'm all right; I bought these burros." But expostulation was useless, as a pair of handcuffs was quickly produced and fastened to his wrists, and chaining him to his saddle, the outfit turned their faces

A Horrible Crime.

toward Gunnison City. The astonished by-standers, upon inquiring the cause of this manœuvre, learned that the man had foully murdered two companions near Ruby City, who were camping and prospecting with him, having shot them both, burying one and leaving the other where he fell. It was supposed the fiend had murdered them for their outfit, burros and money, although very little money was found on his person. The party camped at Ohio City for the night, where the guilty man was in terrible fear of being lynched, which luckily he escaped.

CHAPTER XVI.

THE MINES OF PITKIN.

THE mineral belt of Pitkin is very extensive, and with the town as a centre, a circle of which the radius is ten miles will scarcely cover the ground in which good mines are being daily discovered. In fact, in a northerly direction, the mineral belt can hardly be found broken until the Tin Cup, Gold Cup and Silver Cup mining districts are reached, in the vicinity of Virginia City and Hillerton.

In a southerly direction the same is true, until the camps on the head-waters of Tumichi River are reached, where important discoveries have been made. Both the former and latter places are distant from twelve to sixteen miles. Westerly the Ohio Creek gold region is distant but about eight miles, the distance between being well filled with promising prospects. Easterly to Miller's camp and the tunnel, eight miles distant, rich strikes have been made all the way along.

The formation is granite, porphyry and limestone, and the veins chiefly contacts, with a dip toward the west of about forty-five degrees. For the most part, the veins are parallel, running northerly and southerly, a little west of due north.

In the vicinity of Pitkin there are places where geologically, things are very much mixed, the internal forces of nature at some former time having shaken and tumbled and twisted the rocks, placing one formation above another, and then, again, *vice versa,* in a manner which will undoubtedly puzzle the most scientific. In many places the rocks still show plainly the evidences of very great heat. The country bears every indication of being exceedingly rich in precious metals, the "float" being very rich which is found on the surface, and many leads cropping out boldly from the hillsides. It would be impossible, at the present writing, to give accurately a fair estimate of the future prospects of the camp or of the prospective value and richness of the mines. The camp is new, being still in its infancy, and the mines are very little developed. The deepest shafts or tunnels are less than a hundred feet, and most of them are not more than twenty-five feet in depth; therefore, with so small development, it would be folly to predict positively the future of the section. But it is sufficient to say, that the prospects are remarkably good, and that rich strikes are being made even at the surface, which promises wonderful things for the future, when the mines shall become more fully developed.

Some of the best known mines in the vicinity of

Pitkin are the Fairview, a bonanza in itself; the Silver Islet, Silver Age, Silver King, Chloride King, Forest Queen, Terrible, Red Jacket, Silver Link, Magnolia, Mocking-Bird, etc. There are many others, probably, as good, but not yet as well known.

The Fairview has reached a depth of nearly a hundred feet; has a fine, strong vein; frequently making rich strikes; running sometimes a thousand ounces in silver. The mine employs a large force of men, and development is being pushed rapidly.

The Silver Islet also gives rich assays in silver, but has not reached the depth of the Fairview. It is in the same vicinity.

The Chloride King is a late strike, giving assays near the surface of one thousand three hundred and eighty ounces in silver, and the owners have already refused a large sum for their mine.

The Terrible mine is opened by two shafts, each about twenty feet in depth. They have struck ore showing free gold in abundance, and assaying rich in silver and gold. The owners have been offered, it is said, forty thousand dollars for their find.

The Red Jacket was sold, we were informed, for forty thousand dollars, and is owned by ex-Governor John L. Routt and others. It is opened by an inclined shaft, something over thirty feet deep, and a tunnel has been run in ninety feet from the foot of the hill, to cut the vein at the foot of the shaft.

A tramway is laid in the tunnel, and work has been begun systematically for successful mining. It has a fine, strong vein, from four to six feet wide, of decomposed quartz, stained with oxide of iron, which assays well in silver.

New strikes are being made constantly, and, with the rumors and excitement of a new mining camp, it is impossible to procure accurate information in regard to all the good mines discovered. The assay offices are full of rich specimens of ore from perhaps a hundred different mines, and these are being added to daily. A strike was made about the middle of June, up on the range near the railroad tunnel, from which assays were made of twenty thousand and forty thousand dollars per ton. Other prospects are being discovered there of great promise.

Coal.

A two-foot vein of coal was discovered five or six miles from Pitkin. But nothing definite could be learned as to its value. There is an abundance of coal in Gunnison County, however, in the vicinity of Gunnison City.

The Hot Springs.

About eight miles south of Pitkin there are some hot sulphur and soda springs, which are fast becoming a great resort for invalids, and those suffering with rheumatism and other diseases, who are able to stand the journey thither. There is no road, but a mere trail or path through the woods, which can only be traveled on foot, or in the saddle. Wishing to visit the springs, we started one morning early on horseback, for the purpose of taking a bath in the healing waters. Inquiring the way just before entering the woods, we were told that it was a plain trail; that we could not miss it, etc. We started on a gallop up the hill on a log road, well beaten, which, however, soon came to an end, being continued only by a narrow trail, a mere path, between trees and under overhanging branches, so narrow that the trees brushed our limbs, and so low in places beneath leaning or partially fallen or lodged trees, that we had to dodge our heads beneath them. Up the mountain we climbed, so steep and precipitous that our horse could with difficulty get along, and was compelled to stop often to get breath. Meantime the trail had become less plain, the path more rocky and difficult, and the traces of horses' hoofs less frequent. We now followed the tracks with the greatest difficulty, and finally found we had lost the trail entirely. We dismounted, hitched the horse, and began to search for the trail. After considerable hunting we found blazed trees and a few horse tracks at some distance away, leading on up the mountain. Unhitching the horse, we led him and walked up the steep hillside, keeping a sharp lookout for the path and blazed trees, which we followed with difficulty. Finally, after passing through two or three miles of forest, we reached the summit of the mountain.

Lost in the Woods.

Here the trail appeared to branch, two sets of blazes appearing, one path bearing to the right, and another to the left. However, we did not at first observe this fact, but observing the right-hand blazes, followed them until the path, after starting down the mountain on the other side, came to a sudden end, and no further traces of horses' feet or tracks of any sort could be found. We hitched the horse, and searched in vain for a continuation of the trail. Leaving our horse, we retraced our steps, taking the back track, but could not follow it far, and failed to find where we had come up the mountain. We searched again, but in vain. We were lost. Oh, the dizzy, sickening feeling in our temples, as we realized this fact. We started for the horse; but where was he? We could not find him. We sat down, and tried to think calmly upon the situation, when suddenly it occurred to us that we had along a pocket compass. We produced it with joy, and order began to take the place of chaos in our mind. We knew this: we had been traveling south; we had at this point turned toward the right—south-west—with our horse. Taking the instrument in our hand, we walked toward the south-west, and soon came in sight of our horse. With the aid of our compass we led him back to the summit of the mountain, where we discovered a blazed trail, leading toward the south-east, which we followed until that, too, came suddenly to an end, and no further marks could be seen. We were again compelled to take the back track, leading our horse, and in passing a blazed tree, on which there appeared to be writing, we stopped to read the inscription, which was as follows:

"SAN JUAN BILL'S TRAIL."

Some one had added, in another hand:

"LEADING TO H—LL."

We felt comforted. San Juan Bill's trail had evidently fooled some other party besides ourselves, and "misery" truly "likes company." We experienced much difficulty in finding the path by which we had come up the mountain. But finally taking our compass in our hand, we struck out boldly toward the north, knowing it would take us to Pitkin, paying little heed to trails or blazes, and reached the foot of the mountain and the main road in safety, resolving to make another attempt to reach the Hot Springs at a future time.

Another Attempt.

Two days later we started again for the Hot Springs. Failing to get any one to accompany us, we were compelled to go alone, as before. We, however, got a gentleman acquainted with the route to start us upon the right trail, as we had previously taken the wrong one from the beginning, both roads entering the woods at about the same point. The road proved to be a plain and well-beaten, though very narrow trail, leading up one mountain, down into a deep ravine, across a small stream, up a still higher mountain, and all this time through dense forest and undergrowth, for a stretch of perhaps four or five miles, when we descended again into a deep gulch, treeless and covered with thick grass, which we followed down until it opened out into a large, beautiful park. In passing down the mountain toward this park, we came upon one of the grandest views upon which our eyes ever rested. Grand forests, lovely valleys, beautiful grassy parks, as green as any meadow; bald mountains, snow-capped peaks, all at one glance. It was, indeed, a beautiful sight. We soon entered the park, which is one of the most beautiful valleys in the Rocky Mountains. It is a mile in width by six miles in length, the surrounding hills sloping gently back from a small

Places of Resort.

brook which passes through it. It is covered with thick, luxuriant grass, as green as an eastern meadow, the surface being as smooth as a rolling prairie, on which there is the greatest variety of wild flowers, in such quantities as the eye rarely encounters, which fairly perfumed the air with their fragrance. The valley has evidently been the bed of a lake at some former period, as the marks of surging waters is plainly discernable in places along its sides, and the timber-line on the surrounding hills on all sides only reaches down to a certain point, which has evidently been the water-line in ages past, previous to the time when the outlet finally cut a channel out through the rocks, allowing the waters to escape.

It is a magnificent park, the rich grass growing far up on the surrounding hillsides, which would afford pasturage for thousands of cattle. Some enterprising ranchman has already brought in a small herd. Following down the stream, which was undoubtedly the outlet of the ancient lake, we soon came to the Hot Springs.

These springs burst out along the mountain side for the space of perhaps one hundred and fifty feet in length, in which there is more than a dozen springs or openings, where the water boils up with a sulphurous smell, so hot as to forbid touching it with the hands, except for an instant. The stones, and in fact the whole earth in the vicinity, is crusted with a white sediment, deposited by the waters. Frequently there are stains, resembling that caused by oxide of iron. From the largest spring we fished out pebbles coated with iron pyrites and mineral, plainly giving evidence of how veins of ore may be formed from thermal springs by the action of water. The great heat of the water may be judged by the fact that the few springs mentioned warm the whole brook of clear, cold water, into which they empty, sufficiently for bathing purposes. Just below the springs this stream has been dammed, and the water raised to the depth of about three feet. Over this pond a log bath-house has been erected, about ten by fifteen feet, and into this warm, clear water we

Taking a Warm Bath.

plunged, and took one of the most delightful baths we ever experienced. The water is lukewarm, and is a clear, running stream, with a smooth, gravel bottom; and it seemed such a luxury to bathe within its bosom, that we found it a hardship to leave its embrace. In a log cabin, near the springs, there were several invalids having rheumatism and other affections, who were drinking the waters and bathing daily, claiming that they were receiving great benefit thereby. Competent judges pronounce the healing virtues of these springs equal to those of the Hot Springs of Arkansas, although those are, of course, much larger.

There are, as yet, no conveniences at the springs in the way of bathing-houses or hotel accommodations, but we were told a small hotel would be erected the present summer.

We returned to Pitkin without incident, except that in climbing a very steep hill, while upon our horse, which was a *very lean* one, the saddle suddenly slipped backward, and ere we could loosen our feet from the stirrups we had taken a somersault backward to the ground. When we had recovered our equilibrium, the horse was quietly lifting his feet from within the saddle-girths, leaving us and our saddle behind. Fortunately we escaped without injury, save a slight bruise on the hand, and arrived in Pitkin in due time.

Game, Fish, etc.

A short distance below the Hot Springs there are some large beaver-dams, which are filled with fish. Large quantities of the largest, finest trout are caught, frequently weighing from two to four pounds each. The woods, too, are filled with game, and it is a perfect paradise for the sportsman or nimrod. Fresh deer tracks are abundant, and the woods are filled with the evidences of bear. We, however, saw no game on our trip, except grouse, and a fox, which crossed the trail ahead of us. We put spurs to our horse and gave chase, and would have fired with our navy, had he not disappeared suddenly into the timber. Bears are very fond of the sweet, tender inside bark of the pine, which lies next to the wood, and we saw many places where they had taken off the bark with their teeth for this purpose. Indeed, there are many places here in the Rockies where the woods are filled with these evidences of bear and we have frequently seen where such marks were made during the deep snows of the past winter, when the crust was frozen strong, more than ten feet from the ground, and upon so many trees alike they showed plainly the great depth of the snow.

Other Curiosities.

Another place of resort, which well pays for a visit to the spot, is to a mountain, about five miles north of Pitkin, called Fossil Ridge. The mountain

Fossil Ridge.

was thus named from the great variety of fossils and petrified objects found upon it, and is, indeed, an interesting locality to visit. Here are found petrified snakes, fish, birds, animals, and the greatest variety of interesting objects, which are a constant source of surprise and delight to those searching for and finding the same.

It is probable that to the sportsman and tourist Pitkin and vicinity offers as great inducements for pleasure and adventure, and as grand opportunities for seeing beautiful scenery and securing game and fish, as any point in the Rocky Mountains.

Pitkin very much needs a smelter or reduction works, for her ores. Grading was done for a smelter in the fall of 1879, and it is rumored that it is upon its way to Pitkin, where it will be at once erected; but nothing definite can be learned in regard to it. As soon as reduction works or stamp-mills shall be erected, Pitkin and vicinity will undoubtedly have a period of remarkable prosperity.

CHAPTER XVII.

OHIO CITY—THE MINES OF OHIO CREEK—GAME AND SPECKLED TROUT—BEAVER-DAMS AND HOUSES—A NEW-MADE GRAVE—AN EPITAPH—A HORRIBLE TRAGEDY AT OHIO CITY—SHOOTING AFFRAY—IN CAMP PROSPECTING—PITCHING OUR TENT—AROUND THE CAMP-FIRE—BAKING OUR OWN BREAD—THE "DUTCH OVEN"—GRAND SCENERY—VIEW OF THE UNCOMPALEGRE RANGE—THE "CAMPING-OUT GLORY"—COOL NIGHTS IN THE MOUNTAINS—STRUCK ORE—DREAMS OF SUDDEN WEALTH—A FORTUNATE "GRUB-STAKE"—DESCRIPTION OF MINES—REPORTED CARBONATE STRIKE—FOREST FIRES—NARROW ESCAPE FROM BURNING—MINERS' CABINS DESTROYED.

Ohio City.

ABOUT seven miles south-west of Pitkin the valley of Quartz Creek again widens out to a beautiful grassy park, just at the junction of Ohio Creek, which flows into Quartz Creek from the north. At this point a town was laid out early in the spring, in fact, partly during our visit, and buildings began to spring up as if by magic. The new town was called Ohio City, and within a couple of weeks contained from thirty to fifty log cabins and tents, and a few good frame buildings, and had a good supply of stores, restaurants, saloons, assay offices, real estate offices and blacksmith shops. Passing down Quartz Creek from Pitkin, the timber upon either side of the gulch gradually grows smaller, as we descend, and finally, just before reaching Ohio City, chiefly

BOULDER CANON, COLORADO.

disappears. This fact has been a great impediment to the rapid growth of the town, as even logs for cabins were scarce, and had to be hauled a considerable distance. There were no saw-mills near until the middle of June, and the lumber brought in from Pitkin over a rough road was very expensive, so that, until the saw-mill was erected within a couple of miles, there was very little activity in building. From that time forward building commenced in real earnest. Town lots were selling rapidly at good figures. A Chicago capitalist invested quite heavily in lots, and was about erecting a large hotel and several store buildings. Ohio City is about a thousand feet lower than Pitkin, and is accessible by a good road from Gunnison City all the year. It is beautifully situated, there being splendid water and most excellent fishing and hunting. We saw here in July some of the finest strings of speckled trout that it was ever our good fortune to see. On our first visit to Ohio City we walked from Pitkin, and in passing a lonely spot in the valley, on our way down, perhaps a mile below the latter place, we passed what was evidently

A New-Made Grave.

Stakes marked the spot for head and foot stones, and having heard of the death of a young man in Pitkin two days previous, while far from home among strangers, we solemnly approached the spot to examine the marks upon the stake.

We stooped down and read the following lines, written in a clumsy hand, evidently by some freighter, who had lost a *dear friend:*

> "Death went prospecting,
> And he was no fool,
> Here he struck faithful Pete,
> The emigrant mule."

A Horrible Tragedy.

On the 26th of May a horrible tragedy was enacted in Ohio City, which cost the lives of two men. It appears that some trouble arose the night previous between two men, by the names of Reid and Edwards. Early in the morning, Edwards walked over to the tent where Reid was sleeping, and called to a boy standing near the tent to look out and go away. Reid hearing his antagonist's voice, sprang up and seized a revolver. Edwards fired at Reid, but missed. A regular fusilade of shots then commenced, in which both men fired two shots or more each, when both fell dead on the spot with their heads near together, shot through their hearts. Both were formerly from Leadville, and of bad reputation. This affray, however, was the only thing of the kind to disturb the peace and quiet of the vicinity, and thereafter person and property were as safe in Ohio as in any town in the land.

In Camp, Prospecting.

The reported rich strike of carbonate ores on Ohio Creek created much excitement, and sent hundreds of prospectors thither. We, too, concluded to go with the rush, and rolling up our blankets and a small tent, and packing them on our backs, with frying-pan, cooking utensils, provisions, etc., we marched to Ohio City, a walk of seven miles; thence two miles up the beautiful, dashing stream of the same name, when we went into camp for a time with another party of prospectors. Our party consisted of three, and our neighbors a party of four. We selected a beautiful grassy knoll, on the east side of the creek, beneath the shadows of high mountains on either side of the gulch, and proceeded to pitch our tent. After the tent was in position, we took a blanket, and going to the banks of the stream, along which there was large quantities of long, dry grass, of last year's growth, we soon pulled a blanket full, which we spread in the tent, on the ground, for our bed. Spreading our blankets over this, we had very comfortable quarters for the night. The other party were old mountaineers, and had no tent, but in the meantime had erected a *brush house;* that is, an inclosure covered on top and at the sides with pine boughs, which, when finished, made a very comfortable shelter in which to sleep, as in this locality rain is not frequent, snows being the chief storms. Before we had fairly finished our work, it was snowing on the mountains within plain sight, leaving them white when the squall was over, although this was the 31st of May.

In the evening we all gathered around our large camp-fire, and stories and songs were in order, and the evening passed away very pleasantly. Late in the evening we crawled beneath our blankets, and slept soundly till the sun was shining brightly in the morning. After breakfast, which consisted of fried bacon, bread, of our own baking (without butter), dried beef, coffee and sugar (without milk), and canned apple-butter, we climbed the mountain to prospect. Perhaps we should describe our manner of baking bread, for the benefit of those not accustomed to camp life in the mountains. Nearly every party of prospectors have what they term a "*dutch oven,*" which is nothing more or less than a kettle, with an iron cover fitting tightly over it, and so shaped that it will hold live coals when placed on top of it. The bread is mixed in a dish, with flour, water, salt and baking-powder; these being the only ingredients used. The kettle is heated somewhat, greased inside, and the bread put into it. The cover is put on, and it is set on the coals to bake. Live coals are also heaped upon the cover. When care is taken not to burn it, very light, good bread can be baked in this manner, and such is the bread used by nearly all prospectors in the mountains.

After ascending the mountain above our camp for nearly a thousand feet, the scenery that met our gaze afar off to the south-west was beautiful beyond description. It was a scene to which

THE SNOW-CAPPED ROCKIES.

neither pen or pencil could ever do justice. The valley of the Tumichi River, and farther on of the Gunnison, was spread out before us, lying beyond the mound-shaped foot-hills of the Rockies, and far beyond all the great Uncompahgre Range, in the San Juan region, covered with snow all the year, and the high, sharp peak of the same name, rising to a height of over fourteen thousand feet, with the mountains, valleys, table-lands and cañons in the intervening distance, formed a panorama of such a vast extent of country that it is indescribable.

Our search for mineral this day proved unsuccessful, and we reached camp tired and discouraged, but with appetites sharpened so that our bread and bacon tasted like the greatest luxuries on earth. Who could fitly describe the "camping-out glory" we experienced in this wonder-land in the Rockies! The beautiful stream of clear, cold water, which coursed past our camp, was full of speckled trout. The mountains were filled with minerals and gold, and the grand pine forests were full of game. Fresh deer tracks we encountered everywhere; frequently some of our friends would encounter a drove of six or seven deer. We saw a fine one within easy shooting distance, but had no rifle with us. Some parties in a neighboring camp killed a cinnamon bear, and one evening we had grouse eggs for supper, one of our party having found six fresh eggs. The mountains were filled with prospect-holes, shafts and tunnels, many showing good indications of gold, silver and copper.

Ohio Creek, at this point, was also full of beaver-dams, and there were many very fresh evidences of their recent workings not many weeks since. The weather was very pleasant, in fact, delightful; though the nights were cold, even in June, and frequently we found ice in our water-pail half an inch thick at this late date. One morning we started up to our mine very early, about half past five o'clock. The morning was very cool, and we took with us, as was our custom, a pail of water, to drink while at work. We set the pail down by our prospect-hole, and when wishing a drink shortly after, we found to our astonishment it had frozen over. This was the 10th of June.

We prospected for a week in our camp on Ohio Creek, and after four days' faithful labor, struck a vein of ore assaying a trifle in gold and silver. How we toiled and delved into the rock for two days after we struck ore! With willing hands and great expectations, how hard we worked for the hidden wealth which we were confident would be found a little deeper. When Saturday night came how proudly we marched, with tired and weary feet, yet with light hearts, to Pitkin, carrying with us a little bag of samples of the ores upon which we placed such high hopes and expectations. Ever and anon the boys were exclaiming,

"I wouldn't take five thousand dollars for my share;" "She's worth a million;" "By George, I believe we've struck it;" and like exclamations; and, indeed, our ore looked well, and created no little stir among the cluster of friends who gathered about us when we displayed our specimens. We were disappointed somewhat when our assay only showed four ounces in silver and a trace of gold, yet we had faith in our find, and continued to sink our shaft on the vein. We were encouraged by the finds of other prospectors in the vicinity.

One man sent out a prospector to work for him on a "grub stake," which means that the party furnished the miner with provisions, tools and "grub," and was to receive a half interest in all the mines located by the prospector. The miner was gone two days, when he struck a lead, and the party who sent him was offered one hundred dollars for his interest in the find, and accepted it, making by the operation at least ninety-five dollars clear in two days. Another party, who camped beside us, located a claim near by, worked on it a couple of weeks, and disclosed a four-foot vein, assaying twenty-eight dollars to the ton, and occasionally showing free gold. He sold one-third of his find for one hundred and fifty dollars, and a few days after sold another third for three hundred dollars, or receiving four hundred and fifty dollars cash for about three weeks' work.

The Mines of Ohio Creek.

The formation along Ohio Creek is mostly granite and micaceous slate or schists. There are many parallel veins running with a very regular strike toward the north-west and south-east, with a dip toward the west of from forty-five to sixty degrees.

The Dakota lode was sold to Chicago parties for fifteen thousand dollars. It assays forty-six dollars per ton in free-milling gold ore. The parties who bought it, when approached in regard to selling it, would not offer to bond it for less than one hundred thousand dollars. The Western Hemisphere is also upon Ohio Creek, which is a valuable lead, owned by the same parties as the Dakota.

The Bluebird lode is a strong vein also, being fifty-one feet wide, of free-milling gold ore, opened by tunnel about thirty feet long.

The Ontario lode, which is also on the same lead, and an extension of the Bluebird, is a valuable mine of the same character of ore. The average value of the ores in these two mines is said to be about fifty dollars, while some of it assays much higher.

The Brooklyn Girl, in the same vicinity, is a strong lead of the same character of ore.

The Clara is also a large, wide vein, of decomposed quartz, stained with oxide of iron, and carrying about twenty dollars per ton in gold, which is free-milling ore.

A strike was made by a young miner, about ten miles up Ohio Creek, which was pronounced carbonates by old Leadville prospectors. An assay showed thirteen ounces in silver near the surface. The day after the strike a gentleman named Holmes quietly took his pick and shovel, and prospecting in the vicinity of the carbonate find, followed "float," or "indications," for a great distance, and finally struck his pick in the ground about a mile distant. At a depth of three feet he was rewarded by great quantities of copper ore, full of pyrites in large cubes. The vein proved to be a true fissure, four and a half feet wide, and rich in gold, silver and copper.

It is expected that a stamp-mill will be put into the camp during the autumn. The latest news from the region is that a rich carbonate strike has been made about eight miles from Ohio City, and five miles due west from Ohio Creek, and that genuine Leadville carbonates are found in place, and great excitement is the result. However, we place no very great confidence in the result, though a few mines have undoubtedly been struck which will carry carbonates perhaps in paying quantities.

The latest strike, however, in the vicinity, is the "Little Per Cent." mine, on Quartz Creek, between Ohio City and Pitkin, not far from the main road. It is described by a correspondent as follows:

"Another big strike was made on the 16th of July, by Messrs. R. L. Clark, S. M. Clark, J. H. Moore, G. P. Moore, Odis Clough and T. B. Smith. While at work on the Little Per Cent., about four miles from Pitkin, a blast threw out a large amount of ore, and upon closer examination proved to be literally full of what was supposed to be pyrites of copper or iron. A specimen of the rock was taken to town, where it was shown to the German assayer, who at once pronounced it free gold. An assay followed, showing it to run up to the enormous figure of sixty-two thousand nine hundred and eighty-five dollars and sixty-five cents. Another assay, made at another assay office, brought the amount up to nearly sixty-four thousand dollars. The excitement became intense by this time, and hundreds of men flocked and crowded to the German assay office, to look at the specimen of the ore on exhibition. As soon as the above result became known, an exodus of men took place, and by afternoon the hills in the vicinity were nearly all staked, giving them the appearance of a graveyard. All this time Messrs. Clark and company were quietly working along, not yet knowing of their good fortune, until men came to see their mine and told the news. Mr. Clark at once came to town, took a surveyor back and surveyed the claim. The gold is entirely free, and stands out in cubes and nuggets. The rock is of a dark green, heavy spar, with a blende and a quartz formation. The oldest miners claim never to have seen any mineral in that kind of rock. Only one old English miner reports mines in British Columbia containing the same rock also with gold and silver. The mountains along Quartz and Ohio Creeks are full of veins of this nature, but miners have turned from them as valueless."—*Letter to Denver News.*

Forest Fires.

About the middle of June the forests became very dry, and becoming ignited by the wind from frequent camp-fires, the flames swept over the country in every direction, causing terrible destruction to the timber, and endangering the lives of hundreds of prospectors in the vicinity of Ohio Creek.

Narrow Escape from Fire.

A companion, describing to us his narrow escape from the flames, said they heard a distant roar down the mountain-side, but supposing it to be the wind, did not heed it, until looking up a few minutes later they saw the flames leaping up the mountain but a few hundred feet away, and realized their danger. On came the fierce flames, leaping up to the very tree tops, sometimes a hundred feet high, rushing toward them at the speed of an express-train. The boys instantly leaped from their shaft, and dropping their tools, and throwing away their coats and surplus clothing, ran for their lives. They were nearly surrounded by fire; but while the flames were sweeping up the mountain they ran lengthwise of the hills, to endeavor to flank the fire, as it was useless to attempt to keep ahead of it. After becoming nearly suffocated by smoke, and many times at the point of giving up in despair with fatigue, they finally reached a small stream, and wading into the water wet their faces with their handkerchiefs, and thus escaped. Arriving at their camp, a sorry sight awaited them. Their cabin was in ruins, and their overcoats, blankets, bedding, clothing, in fact everything except that upon their backs was burned. It was, indeed, a serious loss. There were many losses of this sort reported in the vicinity. Many camps were burned, and their occupants lost their all in the way of clothing and bedding, and frequently money contained in clothing, etc.

We spent much of our time while in the mountains along Ohio Creek. It was a beautiful country, a delightful climate, and a dry, healthful atmosphere, with here and there some of the most lovely scenery in the world. It was with many regrets and lingering looks behind that we finally bade it a last good-bye, and turned our steps toward Gunnison City.

CHAPTER XVIII.

GUNNISON CITY, THE COUNTY-SEAT OF GUNNISON COUNTY—THE ROAD FROM PITKIN TO GUNNISON—SITUATION AND POPULATION OF GUNNISON—PRICES OF REAL ESTATE, LUMBER, ETC.—RIVAL TOWNS—THE RAILWAY SOON EXPECTED—THE COAL-FIELDS NORTH—ITS NATURAL ADVANTAGES AND PROSPECTIVE "BOOM."

Gunnison City.

GUNNISON CITY is the county-seat of the county of that name, which is probably the largest in the State, and embraces at least ten thousand square miles of territory. It took its name from the Gunnison River, which was named in honor of Lieutenant Gunnison, of the regular army, who lost his life at the hands of Indians, in that locality, many years ago. Grand rivers, towering mountain peaks, and splendid valleys covered with nutritious grasses, are the characteristics of the country. Probably the finest grazing land of Colorado is embraced within the limits of Gunnison County. Watered by the countless small streams which flow down the gulches of the western slope, and blessed with more frequent rains than the east side, the pastures of Gunnison present a striking contrast for the better to those east of the range, except in the older districts, where there are abundant facilities for irrigating. From Pitkin to Gunnison City the road leads down Quartz Creek to its junction with the Tumichi River, where there is a stage station, called Parlins; thence down the Tumichi River, through a beautiful green and grass-covered valley, to its junction with the Gunnison River, where upon a large, level plain, several miles wide by many miles in length, and surrounded by hills and mountains, in the distance is Gunnison City. The distance from Pitkin is twenty-seven miles, and for the most part it is over a good road. The stage fare is four dollars.

After leaving Ohio City, seven miles below Pitkin, the valley of Quartz Creek widens, and the mountains on either side rise less high, and the timber gradually disappears entirely. The foothills on the western slope present a peculiarly mound-shaped appearance, and there are frequent dome-shaped points of rocks and fantastically-moulded granite ledges, which present a great variety of beautiful scenery along the way.

At Parlins ranch, twelve miles from Gunnison, there is a post-office and a few houses, and the stage changes horses. The whole distance from there to Gunnison the valley is taken up and fenced by ranchmen, and upon the green bottom-land, along the Tumichi, there is to be seen fine herds of cattle, and frequent houses occupied by ranchmen and their families.

Gunnison City is beautifully situated in the

GUNNISON'S BUTTE.

splendid valley which here extends up the Gunnison River for many miles, and is in plain sight of the snowy peaks of the Elk Mountains to the north, which have proved so wonderfully rich in minerals. The city has a population of nine hundred, and is growing very rapidly. The elevation is something over seven thousand feet. The town has large business houses, and all branches of business are well represented. It has a sound banking institution, and has a new court-house, and many very good wooden buildings are in course of erection. Lumber has been scarce and high, and still commands fifty to sixty dollars per thousand. There are a few very good buildings of adobe, or sun-dried brick, which, although dried in the sun, appear to be very solid and make very substantial structures. Real estate and rents are high; single stores on the main street rent as high as one hundred and twenty-five dollars per month, and lots are held at from four hundred to one thousand dollars. Like Ruby Camp, and other towns in Colorado, Gunnison has been cursed with rival town companies. There is the old town of Gunnison, and West Gunnison, which is half a mile west. The valley is so wide that there is room for a city as large as New York, and the consequence is that different parties have located the lands in various directions, and have had the whole cut up into streets and lots, the owners of which and sellers of lots can hardly be excelled in lying for their own town and against all the others. The Denver and South Park Railroad Company are said to be interested in West Gunnison, and will assist that town by building there their depot and shops when their road is completed. At present the old town is ahead, and is much the largest and best business town of the two, and after a struggle has succeeded in getting the new court-house. A few years will probably see both towns connected and built up as one town. The railroad will probably reach there within the year 1881, and as it will undoubtedly be a point from which branches will be extended north, and south, and west to various portions of the country, and especially the coal-fields north, Gunnison City is destined to grow very rapidly, and will soon become a populous town, or, perhaps, a flourishing city. Its elevation is not great, the winters not very severe, and the snow-fall is light; and should the railroad company erect shops, as they will necessarily have to somewhere on the west side of the Rockies, Gunnison will surely get the "boom" she is expecting.

CHAPTER XIX.

RUBY CAMP—THE TOWN OF IRWIN—ALONG THE ROAD—BEAUTIFUL WILD FLOWERS—GRAND SCENERY—CASTLE ROCKS—THE COAL REGION —THE FALLS OF BIG OHIO CREEK—CLIMBING INTO THE ELK MOUNTAINS—SNOW ALONG THE ROAD IN JULY—THE TOWN: ELEVATION, POPULATION—RIVAL TOWNS AGAIN—BUSINESS—PRICES OF LIVING— A LAKE TEN THOUSAND FEET HIGH—A BIT OF HISTORY—THE MINES OF RUBY CAMP—RUBY SILVER ORE—EXTRAORDINARY RICHNESS—THE UTE INDIAN RESERVATION—RICH IN MINERALS—SIXTY MILES INTO THE RESERVATION—LIVING ON VENISON AND GAME—FORTY POUNDS OF TROUT—INDIANS NOT HOSTILE—A TERRIBLE ADVENTURE WITH MOUNTAIN LIONS—EIGHT DAYS WITHOUT FOOD—GOTHIC CITY—THE TOWN OF CRESTED BUTTES.

Ruby Camp.

THE road to Ruby Camp leads up the broad valley of the Gunnison River for two or three miles, then crosses the river on a toll-bridge, and strikes into the beautiful valley of Big Ohio Creek, which empties into the Gunnison. This stream, however, should not be confounded with the Ohio Creek named in a former chapter, as the two streams are more than twenty miles apart, and are in valleys far apart. The valley of Ohio Creek is wide, and covered with luxuriant green grass, and well taken up with ranches and dotted with log cabins, and quite frequently fences to protect the meadows where the ranchmen raise hay. On either side of the stream are beautifully-molded foot-hills, sloping gently back to the high mountains in the distance, on which there is no timber to be seen, all being bare to their tops. All along there is to be seen some lovely views of landscape, mountain peaks and stream. The valley, too, is well stocked with herds of cattle and horses far up into the mountains, and frequent irrigating ditches have been constructed by the ranchmen living along the stream. In the distance can be seen snow-covered mountains, and on every side of us were the most lovely wild flowers and roses in full bloom, in great variety. When about twenty miles from Gunnison, we began to climb into the mountains, which began to be covered with timber, and we came into view of some grand mountain scenery. To the west of us rose castle-rocks, about five miles distant, in the form of some huge ancient castle, on the summit of a high range of mountains, with their corners, and towers, and turrets, and dome-shaped roofs. The resemblance is so striking that one can almost imagine the windows and doors, and evidences of former habitation. We also passed a coal region, lying a little to the east of us, which is about twenty miles from Gunnison City, where a very good quality of bituminous coal is found, which makes excellent coke, there being a five-foot vein. This coal basin is said to be quite extensive, and lies just south-east of Carbon Mountain, which is a very high peak, rising to the height of about twelve thousand feet.

At this point we passed a team of eight yoke of oxen, hauling a large portable boiler and engine, for a saw-mill, to be erected at Ruby Camp. The road soon after became so steep that we were compelled to get out of the hack and walk for a mile or more. At this point the head-waters of Big Ohio Creek takes a leap down the steep mountain of about a thousand feet. It is not a vertical fall, but dashes down at an angle of about seventy-five degrees, foaming and milky-white, through a channel cut in the rocks, presenting a beautiful sight.

Up, up we climbed, until we had snow-banks two feet deep in places on either side of the road, and frequently we picked wild flowers with one hand while we could reach the snow with the other. It was the middle of July, yet there was lots of snow to be seen, and the spring flowers were just coming out into bloom. When the summit of the divide was reached, and we could look off toward Ruby, we saw high, snow-capped mountain-peaks and rocky, craggy-looking summits, raising their bald heads about us on every side.

The road for the last four miles in the vicinity of Ruby is in a horrible condition, the mud being very deep, besides being rocky, and through forests, over roots and stumps, and we much preferred walking to a ride over this fearful road. An eight-hours' ride brought us to Irwin, which is the metropolis and post-office of Ruby Camp, or mining-district, the distance having been thirty miles, and we having stopped at Wilson's ranch for dinner about midway on our journey.

The elevation of Irwin (frequently called Ruby City) is about ten thousand five hundred feet. It is situated in a beautiful valley, between high, rocky, bare-looking peaks, and is surrounded on every side by dense forests of spruce. The Irwin portion is on a hill, sloping toward the south-west, and the main street is full of rocks and stumps, and is very rough and uneven. The town was more of a city of tents than any place we had visited in Gunnison County. Lumber, until recently, had been very scarce and high, and nearly impossible to get, which causes had retarded building very much. However, there were two sawmills in operation, and another was upon the way, and the price of lumber had been reduced to thirty-five dollars per thousand, and there was evidently to be great activity in building very soon, judging by the many foundations being erected.

Everything in the place gave evidence of great thrift and enterprise, and of newness and sudden growth. The oldest portion of the town is called Irwin. The new town, called Ruby, is about one-fourth of a mile south, and considerable rivalry has existed between the two places. However, the whole distance between the two will soon be built up, and the whole is known everywhere by the name of Ruby Camp.

All branches of business seem to be well represented, and large stocks of merchandise of all kinds are displayed upon every side, with the usual large supply of restaurants, saloons and bakeries. It has also a weekly newspaper, called the *Elk Mountain Pilot*.

A hasty count of the place, including both towns, gave a population of one thousand two hundred, and four hundred tents and dwelling-houses, seventy-five business houses and sixty-five unfinished buildings, making a total of over five hundred buildings. Besides these, the surrounding hills and woods contain many tents and cabins, filled with campers and prospectors; and undoubtedly Ruby Camp and vicinity has not less than two thousand people. There will soon be three saw-mills, and grading is being done by the "Good-enough" Mining Company, with a large force of men, for a stamp-mill, part of the machinery for the same being on the ground. Prices of staples at retail were as follows: Flour, eight and a half cents per pound; hams, twenty-two cents; bacon, twenty-one cents; granulated sugar, twenty cents; kerosene oil, one dollar and fifty cents per gallon; hay, nine cents per pound.

The main street of Irwin was crowded with people in the evening, giving evidence of the great number of prospectors in the camp, and of the general thrift of the place. Following up this street toward the west, just over a little hill, we came in sight of a beautiful lake of clear water, covering two or three acres, surrounded by a forest of spruce. Across this lake, from near the water's edge, some huge, bare, chalky-looking peaks rise abruptly to a great height, making a very pretty picture in the sunlight toward evening. Upon these mountains were many large patches, or drifts of snow, and, in fact, on all sides of the town plenty of snow could be seen, though it was the middle of July.

A Bit of History.

Ruby Mining District was duly organized according to law, in July, 1879, and the town of Irwin was surveyed and laid out about the same time. Some time in the May or June previous two or three of the mines containing rich ruby silver ores were discovered. A post-office, named Irwin, was established, and a postmaster appointed in September, 1879, and during that year about forty log cabins were erected, and about fifty persons remained during the winter. The winter was a severe one, and nearly fifty feet of snow fell in all during the season. It was at all times so deep that it entirely covered the log cabins, and the mode of getting out and in the houses was up and down a stairway, shoveled in the snow, from a hole in front of the door. A story is related of parties who went there very early in the spring, that after traveling on snow-shoes the proper distance, and supposing they must be near the place, but observing no houses, they suddenly came upon a man standing by a hole in the snow, and where smoke was rising from a chimney, and inquired of the stranger the way to Ruby Camp. The answer was, "Right here, sir; you're in it." The party, amazed, inquired where the post-office was. "Right over in the next hole, sir;" and, sure enough, right over in the next hole in the snow, completely covered with snow, was the post-office.

The survey of the town-site of Irwin was approved at Washington during the spring of 1880, and it is expected that the patent for the land will soon be issued, making titles to lots and property good.

Mines of Ruby Camp.

Real estate and rents were very high. Lots, twenty-two feet front by one hundred feet deep, were held at four hundred dollars, and not to be had at that. Lots rented at from twenty-five to fifty dollars per month, without buildings upon them. However, prices of living were not extravagant. We got good meals at the restaurants for fifty cents, and put up at a lodging-house similar to the one described in Pitkin, at fifty cents per night. The town lies close to the Ute Indian reservation, being only about a mile east of the line.

Ruby Camp is undoubtedly the bonanza mining camp of Gunnison County. The formation is granite, limestone and porphyry, and the veins are undoubtedly true fissures, carrying very rich ruby silver ores. Several of the mines are sacking and preparing to ship ore. The Forest Queen has a two to three-feet vein of ruby silver ore, which has run by mill-test over two thousand dollars per ton. The mine was once sold, we believe, for forty thousand dollars, but could not be bought for many times that sum at present. It is said that the man who located the Forest Queen, while on his way to Ruby Camp, got stuck in the mud with his ox-team, and agreed to locate a claim for the friend who assisted him through the difficulty. He accordingly located the extension to his claim for his friend, calling it the Ruby King. This mine has a vein about the width of the Forest Queen, and has had a mill-run of about two thousand dollars per ton. It sold for sixty thousand dollars.

The Bullion King has a vein three feet wide, with ore running as high as seven hundred and fifty dollars per ton, and sold for one hundred thousand dollars. The Ruby Chief has a vein thirty inches wide, increasing with depth. The ore runs over twelve hundred dollars per ton, and the mine sold up among the thousands. The Lead Chief has a fifteen-inch vein, carrying eight hundred ounces of silver by mill-run, and sold for sixty thousand dollars. Most of these mines have increased in width and richness with depth. In addition to the above, there are the Last Chance,

R. E. Lee, Little Minnie, and many other very promising prospects. There have also been rich discoveries made on the

Ute Indian Reservation,

which lies west and north-west of Ruby Camp. It is a country rich in minerals and coal deposits, and filled with game and fish. In another year the settlement already effected with the Indians for their removal will have become finally arranged, and this vast, rich section will be opened up for settlement and exploration. During the present season there were not less than fifteen hundred prospectors upon the reservation, yet the ground was hardly half prospected, and but a small portion of its riches unearthed. A friend of ours left Pitkin, and, in company with his partner, passed into the reservation sixty miles, and had been out five weeks, when we accidentally met him in Ruby Camp. They had taken with them a large stock of provisions, bacon, etc., when starting, but found game and fish so plenty that they returned with a large portion of their pork. Their company of six or more persons killed venison so frequently that it became almost their daily living. They went as far as the Grand River, which was swollen high, and they did not attempt to cross, but returned. The streams in that vicinity were filled with salmon trout, and our friend said that it was no uncommon thing to catch ten pounds of fish, and one of their party had caught forty pounds at one time. He had located five claims on the reservation, showing ruby and brittle silver, and in the short space of five weeks had worked out the assessments and had them surveyed and recorded. He had been offered fifteen thousand dollars for his find, and had sold a mere location stake, on which no work had been done, for two hundred dollars cash. Their party had seen but two Indians on their way to the Grand River, and these were very friendly.

A Terrible Adventure with Mountain Lions.

Two prospectors, Thomas French and John Shafer, the latter from Cleveland, Ohio, started on a prospecting tour into the reservation with their burro, on which was packed their outfit, tent and provisions. They journeyed westerly from Ruby Camp, and had reached a point about twenty-five miles distant, and Shafer was somewhat in advance driving the burro, and French was about one hundred feet in the rear, when suddenly, while in a deep ravine heavily timbered, French heard a scream from his companion, and saw to his horror six or seven huge mountain lions in the act of springing upon Shafer, and also the burro, almost tearing them in pieces before his eyes. French beat a hasty retreat and escaped, but before reaching Ruby became lost, and wandered about in the wilderness, starved and suffering, until he was found by a prospector, eight days afterward, in a delirious and dying condition. The prospector took him to camp, where he was kindly cared for, but he remained delirious and unable to give an account of himself for several days. Our friend met the unfortunate man after he had so far recovered as to be able to give the facts as above.

The mountain lion of the Rockies is a large and ferocious beast, of heavier build than the panther, and resembling the African lion, except that it has no mane and beard like the latter. They are very numerous in this section, and are frequently found in pairs of two, or more. They are not considered dangerous when met alone or singly, but seem to be very bold when found in groups.

Gothic City

is about fifteen miles east of Irwin, or Ruby Camp, and is upon a continuation of the same mineral belt which extends eastward. There are some very good mines in the vicinity of Gothic City, and the town is a fair rival for Irwin and the other camps in Gunnison County. The place has a weekly newspaper, called the *Gothic Bonanza*, and one or more smelters will be erected during the present summer.

Crested Buttes

is a small town, situated about eight miles east of Irwin, at the junction of Slate River and Coal

Creek. It has in the vicinity an underlying vein of bituminous coal, four feet thick, which makes excellent coke. It is expected that the fact of having plenty of cheap fuel and coke will make it the great smelting point for the ores of Ruby Camp and Gothic, and, therefore, several smelters are in course of erection in the vicinity, and the town shows great activity in the way of building and lot speculation, etc.

CHAPTER XX.

FROM GUNNISON CITY TO SOUTH ARKANSAS STATION—OVER MARSHALL PASS—RECROSSING THE ROCKIES—DISTANCE AND FARE—FIFTEEN HOURS IN A CROWDED STAGE-COACH—INCIDENTS OF THE JOURNEY—WALKING ACROSS THE SUMMIT—SCENERY—PONCHA SPRINGS—THE TOWN OF SOUTH ARKANSAS—RAPID GROWTH—ON TO DENVER—THE TOWN OF CLEORA—THE "DESERTED VILLAGE"—ENTERING THE GRAND CANON OF THE ARKANSAS—MASSIVE SCENERY—THROUGH THE "ROYAL GORGE"—GIGANTIC WALLS, TWO THOUSAND FEET HIGH—CANON CITY—PUEBLO—SIGHTS ALONG THE WAY—PIKE'S PEAK—MONUMENT ROCKS—SAFE ARRIVAL IN DENVER.

Over Marshall Pass—Recrossing the Continental Divide.

WE entered one of Barlow & Sanderson's Concord coaches, at Gunnison City, one morning at five o'clock, and started on our return journey to Denver. The distance from Gunnison to South Arkansas Station, on the Denver and Rio Grande Railway, is sixty-six miles. It occupies about fifteen hours of continuous riding, and the fare is ten dollars. The stage company change horses five times in the distance, using four horses during the first three changes, and afterward using six. The road for nearly the whole distance on the western slope leads up the beautiful valley of the Tumichi River, and for the most part is in good condition. It is a toll-road, and was built at great labor and expense, and was completed at a recent day. At Parlins Ranch, where the first change is made, we halted for breakfast, and had an excellent meal. The stage was soon filled to overflowing, and we began to experience some of the discomforts of stage-travel in the West. Nine persons were packed inside the narrow box, on three seats, so close together that six persons could only occupy them with comfort; besides there were one or two passengers on the box with the driver. The dust was fearful; great clouds were thrown into the coach by the ponderous hind wheels, which had no way of escape, as in an open vehicle, and the consequence was that we were all soon of one peculiar brown color, very much alike. The frequent change of horses was a great relief, however, giving an opportunity for the passengers to get out and exercise their cramped limbs and rest. There was also an excellent dinner station, where a very good meal was furnished at seventy-five cents. Excellent coffee, rich cream, pure milk to drink, and splendid sweet butter, together with the best of home-made bread, all of home production from the owner's ranch, were luxuries which the passengers all seemed to enjoy. In the afternoon, the valley of the Tumichi, which had been wide, began to narrow up to a cañon between the mountains, and at Wilson's ranch, where six horses are attached, the ascent up the Rockies becomes very steep, and the road becomes a narrow, winding "dug-road" along the side of the precipitous mountain, at places carried hundreds of feet up from the bottom of the ravine below. At such times it requires some nerve to ride behind a span of six, when at some sharp turn in the road the leaders dash around out of sight with a sharp precipice below, and the stage, perhaps, takes a sudden tilt toward the brink. However, the drivers carry a steady hand, and understand well their business, and accidents are rare.

Our progress up the mountain was necessarily slow, and retarded frequently by meeting freight trains, which were descending. The road was narrow, and the passing places were sometimes far apart. During one of these hindrances we walked ahead, and reached the summit of the Rockies somewhat ahead of the coach. It gave us an opportunity for a grand view of the massive scenery which surrounded us. Below us, to the west, the eye could follow the valley of the Tumichi far out through the foot-hills toward the plain, and the winding, zig-zag path, by which we had ascended, could be seen far below us for two or three miles. Heavy green forests reached far up toward the summit on the western slope. To the south of us a gigantic peak, upon which there was still patches of snow, seemed to rise to an immense height; and just to the north, and very near us, another massive peak of great height; and far to the east, even beyond the great valley of the Arkansas, were rugged peaks and irregular mountain ranges, many miles away.

We walked across the narrow divide, which is less than fifty feet in width, and started down the Atlantic slope, when the stage overtook us, and we were soon whirling down the mountains at a rapid pace. The descent is even more abrupt than on the Pacific side, and the stage rattled down the sharp pitches at a pace which is anything but quieting to the nerves of tired passengers. However, an hour's ride brought us in sight of the great valley of the Arkansas, and we rolled into the little town of Poncha Springs. Here there are fifty or more newly-constructed houses, hotels, restaurants and business places, situated on the South Arkansas River, a few miles above its junction with the main branch, and which appear to have been mostly built within a year. The celebrated Poncha Springs are within a mile, and are quite a resort for bathers and health-seekers, as the hot soda and iron waters are considered very beneficial.

Roads lead from this place north to the flourishing mining camps of Arbourville and Maysville, where recent rich discoveries have been made, and towns of considerable importance have sprung up within a year.

Five miles farther on, across the wide valley of the Arkansas, with grand, towering mountains in the distance on every side, we reach and cross the

Arkansas River, and the new town of South Arkansas, which had been laid out and built within the two months previous to our arrival.

The Denver and Rio Grande Railway reached the place in May, 1880, making it their terminus for a time, previous to extending their line, which is now completed on up to Leadville. A town was laid out and built, and being the nearest railroad point to Gunnison City and the mining camps tributary to Marshall Pass, it bids fair to become a permanent town. It is now the natural supply point for a large region of country, both west of the range and the new camps of Maysville and Arbourville, and others on the east side. It has about seven hundred inhabitants, good stores, hotels, a bank, and large forwarding houses, and is a busy town for so young a place. It is astonishing to Eastern people how suddenly these Western towns spring up. Here is a place which within six weeks, where not a building, or tent, or aught but the sandy plain marked the spot, grew to the dimensions of a Western city. On a certain day the Denver and Rio Grande Railway extended their line to this point, making it the terminus, thus leaving the older town of Cleora two miles south "out in the cold." A platform was built for the accommodation of passengers and freight, and by the first trains large stocks of goods and merchandise were brought in and placed under canvas roofs, and large sales began from tents. A stampede from the old town of Cleora was the result; stores, saloons and restaurants straightway removed to the new prospective city. Speculation in lots and real estate became wonderfully excited, and soon the old town of Cleora became nearly deserted.

South Arkansas became the head-quarters for Barlow & Sanderson's stage lines to Gunnison, and to Saguache and Lake City, in the south-west, and also became a regular meal station for all through trains on the railway to and from Leadville. Thus the town soon acquired its present bustle and activity.

It was nearly dark when the stage deposited us at the depot, dusty and weary, and we determined to stay over night and view

The Grand Canon of the Arkansas

by daylight. Accordingly we waited, with good hotel accommodations, until eight o'clock the following morning, when we took the cars for Denver. The railway follows down the Arkansas River, which is a very rapid and swift stream. At this season it is swollen and very muddy, from melting snows far up in the mountains. A couple of miles down from South Arkansas we pass the town of Cleora, which truly reminds one of Oliver Goldsmith's deserted village:

> "As now the sounds of population fail,
> No cheerful murmurs fluctuate in the gale;
> No busy steps the grass-grown foot way tread,
> But all the bloomy flush of life is fled."

GRAND CANON OF THE ARKANSAS, COL.

Its former greatness has, indeed, vanished, and empty buildings chiefly mark the spot. However, there are a few residents still, and the train makes a stop at the once busy depot.

Soon after leaving Cleora, the wide valley suddenly narrows up to a ravine, and we enter the head of the Grand Cañon of the Arkansas. For the first few miles going south the cañon frequently widens out in places, forming little narrow valleys, within the rugged mountains, in which there are often a few cabins and habitations, and occasionally a station, at which the train stops. Sometimes, in these narrow valleys, we passed a splendid garden, or a fine crop of corn or vegetables, where the ranchman has irrigated successfully, and these look, indeed, like an oasis amid the vast desert of rocks which surround them. From some of these little valleys some grand, sharp peaks can be seen toward the south-west, rising fairly above the clouds. The misty vapor can be seen hanging about their heads, while just above the vapor the peak is seen like a little mound resting upon the cloud. The river below us is frequently filled with huge granite boulders, and the rocks above overhang the track, and we pass beneath. The gulch keeps narrowing, until huge granite rocks of every conceivable shape rise up abruptly on either side, and there is but the narrow track over which we pass and the river below, hardly averaging sixty feet wide, a swift, surging, rapid stream. The rocks on either side grow higher and higher, and more vertical. Huge granite pillars, of many fantastic shapes, in which the bright spots of mica glisten in the sunlight, and frequent caves and grottoes, of various sorts, are the characteristics for nearly twenty miles down through the cañon. Just before reaching the Royal Gorge the train halts, and an open observation-car is annexed to the rear of the train, in which the passengers who choose are invited to take seats. The view from this car we shall never forget! The grade is steep, and the cars run apparently at a fearful rate of speed. The river below is full of rapids, and is a milky, raging torrent. The road is so crooked and winding that the engine is constantly in sight from one side or the other. The sudden curves, the high rate of speed, the torrent beneath, the jar and hollow rumble of the rushing train, the echoes shut in by the vertical walls, all tend to thrill the pulse and make the nerves tingle with a sort of fear; while the massive walls of granite, which narrow up to from twelve to twenty feet, and rise to the height of more than two thousand feet, form a scene which is a constant wonder and delight. It was nearly noon, yet there were shadows in the Royal Gorge. Nature seemed here to be set on its edge. The gigantic walls of vertical strata, rising in places to most two thousand feet, have a grandeur about them hard to describe. It was our final triumph! We had seen much of Colorado, but naught had inspired us like this. Naught had brought this feeling to our heart, often produced by the chords of a beautiful piece of music, or the sweet strains of a song which touches a tender place in the heart. So this grand picture before us seemed to thrill us with a feeling of awe at nature's handiwork.

At one point, where the walls of the cañon close near together, the railway has built an iron bridge, suspended by wrought-iron braces from wall to wall, running the track over the water by this means for a hundred feet or more, instead of blasting out a roadway in the vertical walls. Soon after passing this bridge the cañon widens, and we soon come out into a very large valley or plain, and enter Cañon City. In this valley we found corn and oats growing luxuriantly wherever irrigation can be practiced.

Five hours from South Arkansas Station brought us to Pueblo, where the train makes a long stop for dinner. Here the main line of the Denver and Rio Grande road, which now reaches into New Mexico, connects with the Leadville branch. The Atchison, Topeka and Santa Fe Railway also runs south and east from Pueblo. North of Pueblo, toward Denver, for many miles much of the country appeared like a barren desert, parched and dry, and bore very poor comparison to the rich valleys and green pastures of the Pacific Slope.

Most assuredly the western slope appears far ahead of the eastern as a grazing country, except in the sections where water is plenty for irrigation.

For a time, in the vicinity of Pueblo, we had nearly passed out of sight of the Rockies on to the rolling desert plain. In the vicinity of Colorado Springs we again approached the mountains, and it seemed a most refreshing sight, after seeing for a while naught but the desert plain.

Colorado Springs is said to be one of the prettiest towns in the State. But from what it takes its name is hard to imagine, as there are no mineral springs or anything of the sort there. Pike's Peak rises up grandly in plain view to the west, and a telegraph line connects the United States Signal Station, on its summit, with Colorado Springs. There on its summit, in a little stone office, three officers of the army are compelled to stay, summer and winter. It is a lonely, tedious task, especially in winter, as it is impossible to get down for months during cold weather, and supplies have to be taken up in the fall sufficient to last the season.

Near Monument Station, about fifty miles south of Denver, there is a very pretty sight from the cars a short distance to the east of the track. A series of rocks rise abruptly from some low hills, which are as regularly moulded as if by a sculptor's chisel, and resemble the pillars and walls of some ancient temple, which has been chiseled by an artist to perfection. They are, indeed, a very beautiful sight, situated as they are among a grove of pines which surrounds them. Near them, just north, is a pretty little lake, on which we saw boats, and ladies rowing. All is in plain view from the train.

The last fifty or sixty miles toward Denver the country became richer, the grass more green, and a great improvement was noticed over the sands of Pueblo and vicinity. Large irrigating ditches bring down a good supply of water, and the corn and oat crops were looking very fine. We arrived in Denver without accident about dark, having been twelve hours out from South Arkansas Station.

CHAPTER XXI.

LEADVILLE, THE CARBONATE CAMP—EARLY HISTORY—CALIFORNIA GULCH IN 1860—FORMER RICH YIELD FROM THE PLACER MINES—THE "HEAVY SAND" THAT TROUBLED THE SLUICE-BOXES—FOUND TO BE CARBONATE ORE—THE FIRST MINES LOCATED—THE FIRST SAMPLING WORKS—THE FIRST SMELTER—A STORE IN JUNE, 1877: BY WHOM ESTABLISHED—THE FIRST BUILDINGS IN LEADVILLE—MARVELOUS GROWTH IN 1878—POOR MEN RAISED TO SUDDEN WEALTH—HIGH PRICES OF REAL ESTATE—RENTS—BUSINESS—SMELTERS—ONE MILLION DOLLARS PER MONTH—A FEW OF THE MINES—VALUE OF THE ORES—THE LITTLE BONANZA: TRIANGLE—A FEW OF THE MEN WHO STRUCK IT RICH—FIVE MILLIONS IN PROFITS FROM MINE SPECULATIONS—SOME OF THE BIG BONANZAS—THE R. E. LEE MINE—THE CHRYSOLITE—THE MORNING STAR—THE LITTLE CHIEF—THE LITTLE PITTSBURG—COST OF LIVING—THE GREAT STRIKE FOR WAGES.

Leadville, the Carbonate Camp—California Gulch in 1860.

SOME time during the summer of 1860, a few hardy prospectors entered the valley now known as California Gulch. Through this valley flowed a small stream, along whose banks skirted a heavy growth of timber. Panning out some of the gravel, it was found to be rich in gold. The reports that went out from this strike soon made the gulch alive with men, and within a few months, it is said, five thousand men had gathered along this little stream up among the mountains. A little town of cabins and log huts sprung up, and was christened by the name of

BUSINESS STREET OF LEADVILLE.

Oro. The gulch proved to be immensely rich, and some claims soon yielded as much as one thousand dollars per day, and it is stated that one firm took out one hundred thousand dollars in two months. The "rocker" and sluice-box were put in operation from one end of the gulch to the other for a distance of six miles, and some claims yielded an ounce or two of gold per day to the man, and one single pan of dirt is said to have yielded five ounces of the precious metal. The gold dust, as it came from the sluices, passed at a valuation of about eighteen dollars per ounce, and in every place of business there was a little pair of scales for weighing it, and gold dust was the medium of exchange. In 1861 the camp saw its best days, and from that time began to decline, as the richest ground became worked out.

For three or four years more it continued to produce considerable amounts, and before the close of 1865 had produced in all over four millions of dollars in gold. The subsequent years, up to 1868, brought forth about seventy-five thousand dollars more, but by that time the camp was about deserted. In 1868, the "Printer Boy" gold mine was discovered. This proved to be valuable enough, so that a stamp-mill was erected to crush its ore. This mill, in the next six years, produced about two hundred and fifty thousand dollars in bullion, but was not very successfully managed thereafter. In 1874, W. H. Stevens and

A. B. Wood came into California Gulch, and began to construct a long ditch to work the placer claims along the stream, by hydraulic process, and to bring in water sufficient to wash the unworked gravel-banks higher up, that bordered the stream. They brought the water from the head-waters of the Arkansas River, and two or three years were required to complete the enterprise, and the first full summer's work was not put in until that of 1878. During the time since work first began in

The Heavy Sand in the Sluice-Boxes.

the gulch, miners had been troubled with a kind of heavy sand, that filled the sluice-boxes, but, of course, deemed it of no value. Messrs. Stevens and Wood had assays made, and found it to be carbonate of lead, carrying considerable silver.

Others made similar discoveries about the same time; this was in 1876–77. It seems, however, that all were very quiet about making their discoveries known, until they were in a fair way to secure Government titles to their claims.

Stevens and Wood located mine claims, the principal ones being the Iron, the Dome, Rock, Stone, Lime, and Bull's-eye. The first "strike" of note was made by the Gallagher Brothers during the same fall and winter, and was called the "Camp Bird." It was near the Iron mine, but farther toward Stray Horse Gulch. In 1877, these mines began to produce ore enough, so that in

April of that year A. R. Meyer, purchasing agent for the St. Louis Smelting Company, established sampling works for the purchase of silver ore, near the present site of Leadville. A month later the St. Louis company began to erect a smelter and blast-furnace, and had it in operation early in the fall.

In June, 1877, Charles Mater started the first building in Leadville, and opened with a stock of groceries. Several other cabins were completed by the time Mater began to sell goods.

H. A. W. Tabor, who had conducted a small store for many years up the gulch at Oro, and later at Fairplay, brought a stock of goods to Leadville soon after Mater had established his store. Thus began the city which to-day, less than three years later, after having passed through such a mining excitement as the world had never seen, boasts of a population of over twenty thousand people. The history of Leadville has been so frequently "written up," that we shall not endeavor to give in detail all its subsequent history since 1877. We shall, therefore, give only such a portion of its history as will interest the general reader.

It was not until 1878 that any considerable excitement was manifested. Since that time there has been a constant stream of immigration pouring into the city. At one time it seemed as if the whole world was rushing to Leadville. Hotels, restaurants and lodging-houses were hardly half sufficient to feed and lodge the army of strangers who were arriving daily. The few hundred who found the camp as early as January, 1878, became as many thousand before the summer was half gone. A city government was organized, officials elected, newspapers established, hotels and banks and business houses built and opened, all within an incredible short time.

In March, 1878, the first large sale of mining property was made. The St. Louis Smelting Company purchased the "Camp Bird" mine, and some adjoining claims, for the snug sum of two hundred and twenty-five thousand dollars. This sent the camp "booming." Those who sold had been poor, hard-working men previous to this time, and their rise to sudden wealth began to spread the fame of the new camp far and wide. Mines were located by the thousand. On some of these shafts were sunk, and the carbonate strata being reached, the fortunate owners were able to sell their claims for fabulous sums. On some of them the outlay had been small, and, of course, the profits from the sale were immense. Men who had previously scarcely ever had a dollar ahead, found themselves suddenly "rolling in wealth." The excitement became so great that the whole country was supposed to be underlaid with the rich carbonate strata, which lay flat, as was supposed, like a coal measure. Accordingly, any kind of a prospect-hole at one time could find a ready sale, even though it had in it not a particle of mineral. Its location would sell it for a good round sum if it were anywhere near the producing mines. Some men from a small investment in a barren prospect-hole, upon sinking their shafts, struck the ore body and made handsome fortunes. A German, from an investment of thirty dollars in the Little Chief mine, is said to have realized sixty thousand. Others, less fortunate, lost their investment. It was found the carbonate belt had limits, and that it was not everywhere of the same richness. Still the mad rush to Leadville continued Capitalists and speculators flocked to the new El Dorado, as well as the mechanic and laborer. One hundred men per day was the average arrivals in the city for much of the time during the first six months of the year 1878.

High Prices of Real Estate and Rents.

Speculation ran rife in everything. Town lots, bought for a trifle in the year previous, sold frequently for thousands each in 1879. The land on Chestnut Street, bought formerly from the Government at two dollars and a half per acre, sold readily a year later for ten thousand dollars for a seventy-five foot front. The purchaser of such a lot sold fifty feet of it the very next day for the same amount. The Grand Hotel, built on the same street at a cost of eight thousand dollars, was sold in a few months for fourteen thousand, and, later, could not be purchased for forty thousand. Nine lots were sold in the suburbs for one thousand five hundred dollars. Within twenty-four hours the party who bought them was offered three thousand dollars for his bargain. Lots on Chestnut and Harrison Streets, which sold in 1879 at one dollar per foot front, were eagerly taken early the next year at one hundred and fifty to two thousand dollars per foot. The Theatre Comique, a one-story frame shell, fifty by one hundred feet, rents for the enormous sum of one thousand seven hundred dollars per month, or over twenty thousand dollars per year; and the receipts here have run up to one thousand two hundred dollars per night. A log house, opposite, rented for seven hundred dollars per month. The City Hall building, on Chestnut Street, by no means an ornate affair, rents for seven thousand two hundred dollars per annum.

In some respects, Leadville is the most remarkable city the world has ever seen, and it probably came nearer being built in a day than any city of its size in the universe. It is finely situated on the left bank of California Gulch, on a broad and gentle slope, near the foot of the Mosquito Range, in the midst of the Rockies, surrounded by towering mountain peaks. The streets lying parallel to the general direction of the mountains are comparatively level, having everywhere an easy grade.

Its elevation is ten thousand three hundred feet, while some of the surrounding mountains rise grandly to fourteen thousand five hundred feet, their summits generally capped with snow.

The growth of Leadville is unprecedented. In January, 1878, the camp consisted of about twenty log cabins. Four months later, the number of buildings had increased to four hundred, and at the present time there is not less than two thousand buildings of various sorts, many being fine structures of brick and stone. In June, 1878, its population was fifteen hundred. In January, 1879, a census taken gave a population of over five thousand. Since then it has grown past all precedent, and in January, 1880, was variously estimated at from twenty to thirty thousand, and even higher.

Every branch of business is represented in Leadville, from the wholesale house, doing a large trade with the surrounding mining camps, to the apple and orange vendor on the street. The streets are crowded with men, and teams, and vehicles, and vim, and bustle, and business are the marked characteristics of this wonderful city. The city has good water-works, an excellent police department, a fair fire department, and has lately put in a fire-alarm and gas-works. It has also a telephone exchange, connecting all the principal business places of the city with a general office. All this, and much more, has been accomplished in the short space of a year and a half.

Thirty-four smelters are reducing ore from over two hundred and forty producing mines, turning out, when all in full blast, nearly one million dollars in value per month.

Some of the Mines—Value of Ore, etc.

Some of the best-known mines, with the value of the ore per ton, will be shown in the following statement of the average amount of silver and lead per ton. It should be borne in mind, however, that no perfectly reliable table can be made in so new and active a mining camp as Leadville. It is correct only so far as it goes, as there are still many other mines which are yielding almost as profitable results, and more are being constantly discovered:

Name of Mine.	Ounces Silver per ton.	
Agassiz, average.............................	35	20 per cent. lead.
Adelaide, average...........................	30	50 " " "
Carbonate, average, 1st grade..........	600	
" " 2d "	200	
Crescent, average, 1st grade............	200	
" " 2d "	80	
Colorado Chief, average..................	24	20 per cent. lead.
Carboniferous, average.....................	125	
Chrysolite, average...........................	125	
Cyclops, average..............................	80	
Dyer, average, 1st grade.................	375	
" " 2d "	125	
Double-Decker, average, 1st grade...	196	and 5 ounces gold.
Evening Star, average, 1st grade......	170	40 per cent. lead.
Gone Abroad, average, 1st grade.....	296	
" " 2d "	100	
Henrietta, average...........................	46	44 per cent. lead.

Name of Mine.				Ounces Silver per ton.		
Little Pittsburgh, 1st grade			150	"	20 per cent. lead.	
"	"	2d "	80	"		
"	"	2d shaft, 1st grade,	200	"	65 per cent. lead.	
"	"	" 2d "	80	"		
Little Chief, average			120	"	30 per cent. lead.	
Lima, average			15	"	24	" "
Morning Star, 1st grade			180	"	40	" "
"	" 2d "		75	"		
New Discovery, average			160	"	30 per cent. lead.	
North Star, average			80	"		
Silver Wave, 1st grade			290	"		
"	" 2d "		100	"	45 per cent. lead.	
Terrible, average			25	"	40	" "
Vulture, 1st grade			238	"	68	" "
"	2d "		80	"	65	" "
Iron, average			225	"	45	" "

Probably one of the richest little finds ever made in this wonderful camp was the Triangle mine, a little patch of ground, triangular in shape, whose side-lines were only thirty-four feet in length. It is surrounded by the "Little Chief," "Chrysolite," Vulture" and "Little Eva," and had been overlooked. It is said that a chain-man, while assisting in the survey of one of those claims, discovered the little vacant lot, and immediately dropped his chain and located the ground. It was a lucky find for him, for from this limited space over fifty-eight thousand dollars were taken out at a cost of less than five thousand dollars. The mineral is now exhausted, unless a second deposit exists below. How many other triangles are nestling in and around Leadville will not be known until after years of further prospecting.

A List of a Few who "Struck it Rich."

The subjoined list ranges from men who reached Leadville without a dollar, to those who had capital to begin operations, and does not include in full the sales made by each or all interests yet held by them, but only such transactions as are known to the public. The list would be greatly extended by searching the records, and by mention of the army of successful business men of Leadville:*

Dick and Pat Dillon made $150,000 from the sale of "Little Chief."

George Spencer, $6,000 from sale of interest in the Chrysolite.

Jack Calhoun, $17,000 from sale of interest in Black Prince.

Mike Morris, $40,000 in "Wolf Tone" mine.

A. B. Wood, $40,000 in Iron mine.

J. C. Langhorne, $62,500 in Vulture mine.

J. W. Johnson, $62,500 in Vulture mine.

Al Rennic, $62,500 in Vulture mine.

R. M. Moore, $9,000 in Vulture mine.

W. B. Page, $9,000 in Vulture mine.

Breck & Co., $11,500 in Vulture mine.

C. B. Rustin, $6,000 in Vulture mine.

John H. Talbut, $18,000 in Vulture mine.

Geo. H. Fryer, $40,000 in New Discovery mine.

Charles, Pat and John Gallagher, $250,000 in Camp Bird and other mines.

---

* For authority, see New Year's edition, Denver *Tribune*, 1880.

A. P. Moore, $25,000 in Matchless mine.

T. Benton Wilgus, $75,000 in Matchless and other properties.

John Borden, $25,000 in Chrysolite and other mines.

Jerome B. Chaffee, $250,000 in New Discovery.

August Rische, $262,500 in Little Pittsburg.

H. A. W. Tabor, $1,300,000 in Little Pittsburg Consolidation.

James Healy, Mike Mackey, Patrick Nash and Michael Brown, $35,000 from Colorado Prince.

Howard Oviatt, George Washburn, T. J. Cooper and Peter Klinefelter, $62,000 each from Scooper mine.

John Borden, Jr., $40,000, in New Discovery.

E. C. Kavanaugh, C. Visscher, D. Rainey, W. K. Burchinell, Charles and Peter Peterson, and —— Meek, $195,000 in Denver City mine.

J. T. Monroe and George Williams, $50,000 in Little Eva.

Captain Jacque, $175,000 in Smuggler mine.

Frank Caley, $50,000 in Undine mine.

Eddy & James, $50,000 in Robt. E. Lee mine.

James W. Younger, $6,000 for one-eighth interest in the Deer Lodge mine.

George T. Hook, $140,000 from Little Pittsburg mine and ore.

Nelson Hallock and Captain Albert Cooper, $250,000 from Carbonate mine.

Captain Plummer, $300,000 in Yankee Doodle.

Breece Estate, $75,000 from Breece mine.

Tim Foley, $97,000 in Matchless mine, $25,000 in Union Emma mine, and owns $200,000 stock in the Highland Chief Consolidation, besides other valuable property.

George W. Trimble, $100,000 in Winnemuc mine, and is largely interested in Highland Chief and other mines.

Jed H. Bascom, $25,000 in Union Emma, and is part owner in Highland Chief Consolidation.

Charles W. Tankersley made $50,000 out of the Highland Chief Consolidation, and is owner in other valuable properties.

Henry W. Wolcott made $115,000 on sale of the Robert E. Lee.

James V. Dexter realized $20,000 from R. E. Lee, and has mining interests worth $60,000.

James Y. Marshall has an interest in the R. E. Lee worth $200,000.

J. S. Fitz made $100,000 on sale of Little Chief and other mines.

Jacob Saunders, William Parker and Colonel R L. Hopkins divided $150,000 on sale of Small Hopes Mining Pool property.

William H. Bush has made $200,000 by real estate.

S. H. Foss, $125,000 in Winnemuc, and is heavily interested in Highland Chief and others.

Ex-Governor Routt, George C. Corning and James Watson are rich by reason of ownership of the Morning Star mine. A grand total of nearly five millions of profits in mine speculation in a single camp.

Some of the Big Bonanzas.

Probably one of the richest mines in the world is a small claim, located on Fryer Hill, near Leadville, called the Robert E. Lee. Early in 1879 a gentleman in Denver was offered an interest in the mine for the modest sum of six hundred dollars, but had no confidence in the claim, and never gave it a thought afterward, until late in the season, when a very rich strike was made in the mine. Now the mine is one of unparalleled richness, turning out some of the richest ore known in the history of mining. On January 14th, 1880, one hundred and eighteen thousand dollars' worth of ore was taken from the mine in eighteen hours, some of it rating as high as eighteen thousand dollars per ton. From Monday, January 4th, to Tuesday, January 13th, the production had exceeded ten thousand dollars per day. This fact induced the owners to make an effort on some particular day to see how much of the valuable ore could be mined and raised. Accordingly, at one o'clock P. M., the start was made for the big run. The ore was accordingly kept by itself, and shipped to Eddy, James & Co.'s sampling works, where it was assorted and carefully tested and sold. Five of the lots assayed as follows: Lot No. 1, 11,830 ounces per ton; Lot No. 2, 4,993 ounces per ton;

Lot No. 3, 1,234 ounces per ton; Lot No. 4, 1,088 ounces per ton; Lot No. 5, 568 ounces per ton. The whole production of the eighteen hours was ninety-five tons, which averaged over a thousand ounces per ton, bringing about eighteen thousand five hundred dollars. For several months thereafter the average daily output of the Lee mine was forty tons of rich ore, though, of course, not always as rich as the above.

The Chrysolite Mine.

During the early part of 1879, this claim was regarded as only valuable because of its proximity to the Little Pittsburg and Little Chief mines, on Fryer Hill. It is now considered as one of the best mines in Leadville. The product during the month of January, 1880, was one hundred and fifty-five thousand three hundred and sixty-four dollars and twenty-three cents; and during one week, ending January 17th, one thousand one hundred and thirty-five tons of ore were extracted, for which the company received the snug sum of ninety-nine thousand five hundred and seventy-three dollars and seventy-one cents, or at the rate of one thousand two hundred and twenty-four dollars and eighty cents per day. And its record ever since has been extraordinarily good.

The Morning Star Mine,

owned by ex-Governor J. L. Routt and others, has a high reputation, and is fast making its owners rich. A piece of selected ore, chloride of silver, was taken from the mine early in 1880, which ran as high as thirteen thousand eight hundred and eighty-four ounces of silver per ton, or at the rate of fifteen thousand six hundred and eighty-nine dollars and ninety-two cents. This was not, of course, a fair sample of the whole ore body, as it averages only about fifty ounces in silver and fifty per cent. in lead. The great value of the mine consists in its large ore bodies and great percentage of lead, an element in the reduction of ore of the utmost importance, having its own flux for smelting, for without that no smelting can be done.

The Little Chief Mine.

This claim turns out a large amount of rich ore. The ore output for the year 1879 was over seventeen thousand tons. But within the time much dead-work had to be done to get the mine in shape for heavy production. The mine cost the Chicago company who own it three hundred thousand dollars. They have, it is said, realized from it, over and above expenses, five millions, and then sold one-half of it for one million seven hundred and fifty thousand dollars.

The Little Pittsburg Consolidation of Mines.

This company have, in less than two years since they were organized, paid in dividends one million and fifty thousand dollars; eight hundred and fifty thousand dollars of it having been paid up to January 1st, 1880. The total product of the mine up to June 1st, 1880, was thirty thousand eight hundred and four tons of mineral, for which the company received one million five hundred and ninety-one thousand and thirty-eight dollars and seventy-two cents. The expenses of extracting and selling it, including office expenses, were five hundred and fifty thousand and sixty-four dollars and eight cents, according to an official statement made July 1st. The company had a financial record almost unprecedented in the annals of mining up to the spring of 1880, when, through the mismanagement of officers, stock manipulations by tricky directors, and the unnecessary exhaustion of ore reserves, the company had to stop paying dividends and fell into bad repute.

Many other valuable mines might be noticed in this connection, but we have not space for an extended sketch of all the rich bonanzas of Leadville, and must draw a line *somewhere;* so we omit them.

The Cost of Living, Prices of Provisions, etc.

The cost of living in Leadville, during 1879, was not high in proportion to most new towns and mining camps, and now that the railway has reached the city, prices will be still further reduced. Flour was five to six dollars per hundred weight; potatoes, five cents per pound; butter, forty cents; eggs, forty cents per dozen; sugar (granulated), fifteen cents per pound; hams, fifteen cents per pound; bacon, fourteen cents; fresh meats, thirteen to twenty-five cents; kerosene oil, seventy cents per gallon; syrup, one dollar and a half per gallon; lard, fifteen cents; fresh milk, twenty cents per quart; hay, one hundred and twenty dollars per ton; oats, six cents per pound; lumber, fifty dollars per thousand; board, seven to ten dollars per week; doctors' visits, three dollars; saddle-horses, per day, three dollars and a half; first-class hotels, three to four dollars per day for transient custom.

The Miners' Great Strike for Wages.

About the first of June, 1880, the miners of Leadville struck for higher wages; or, as some claim, against a reduction of wages. For a week or more business was at a standstill. Most of the smelters had to blow out, and furnaces had to be allowed to cool from which the fires were never out before since they started. The miners, in a mob, marched to the various mines and smelters, and with threats of violence prevented men from going to work. In consequence, Leadville's production of thirty thousand dollars per day or more suddenly decreased to mere nothing. It was a severe blow to the city, from which her merchants and business men have as yet hardly recovered. Troops were called out to preserve order and guard property from destruction; the governor declared martial law, and for a time things assumed a serious look. Frequent threats were made by miners and roughs of burning the city, but no such attempt was made.

This condition of things lasted for some time, but at length order was restored. The mine owners, in some cases, under protection of the military, filled the places of strikers with new men and resumed work. Finally the labor difficulties were arranged, and one after another of the mines and furnaces started up, until all were in full blast. The riot, however, left its effects behind, and it proved the cause of driving from the city to other mining districts many miners and a large floating population; and it is said that Leadville has since been dull, and hardly boasts of its former greatness.

CHAPTER XXII.

*STOCK GAMBLING—MINING SPECULATIONS—TRICKS OF SHARPERS, ETC.— THE ASSESSMENT LAWS OF CALIFORNIA AND NEVADA—"FREEZING OUT" SMALL STOCKHOLDERS—COMSTOCK MANAGEMENT—THE CONSOLIDATED VIRGINIA AND CALIFORNIA MANAGEMENT—ASSESSMENTS LEVIED IN NEVADA IN 1879—DIVIDENDS OF THE UNITED STATES IN 1879—THE LITTLE PITTSBURG STOCK BUBBLE—GREAT DECLINE IN VALUE—A COMMON SWINDLE—"WILD-CAT" MINES—TUNNEL SCHEMES —REPORTS OF PROFESSED EXPERTS, ETC.—BIG PROFITS TO PROMOTERS IN STOCKING MINES, ETC.—"SALTED" MINES, ETC.

Stock Gambling, Mining Speculations, Tricks of Sharpers, etc.

GOLD and silver mining being a business which requires a large expenditure of money and heavy investments of capital, often for a considerable length of time before returns or profits can be expected, it very naturally happened that the operation of mines by means of stock companies and large corporations became the favorite method among capitalists, as the risk is very great for private enterprise. Undoubtedly the organization of stock companies, when properly officered and honestly managed, is the proper method of operating mines. And there is no question, but that the *bona-fide*, actual sale or purchase of mining stock, carried on with honest intent, is as legitimate a business as any other. But, like many other branches of business, such as the speculation in railway shares, and dealings in grain and produce at the exchange on "margins," it frequently happens that rogues occupying the positions of trusted officials, and that smart operators at the exchanges, find an opportunity to manipulate the price of the stock of a really valuable mine to suit their own dishonest purposes, often to their own enrichment and to the loss and ruin of the unknowing stockholders. Fortunes have been made by such men in this manner, and are still being made. There is a large class of operators in mines who care nothing whatever for the profits that may accrue to them from legitimate mining, or the production and profits from ore extracted from the mine or mines which they may control. Such is too slow a method of accumulation for them; they seek only the profits which may come from their manipulation of the prices of the stocks they own, and the rise and fall of the market for the shares of the mines which they control. Until recently, San Francisco has been the point where the greatest operations in mining shares have taken place, and the speculations there were upon a gigantic scale. The frauds that have been practiced upon innocent shareholders, by shrewd managers of mines on the Pacific Coast, have been immense. The laws of California and Nevada, allowing the officers of corporations to call assessments on mining shares, greatly aided the operators in "fleecing" small stockholders, and in defrauding them of their rights.

The law allowing the privilege of assessing shares is not a bad law, when honest managers of mines call *only legitimate assessments* for actual amounts required to carry on the business of mining. Frequently the assessment plan has advantages over the non-assessment plan, and is preferable, as in non-assessable shares, frequently the officers have to borrow money, and mortgage the company's property, to raise money to carry on developments upon the property; while a small assessment upon each share, honestly and legally enacted, would have kept the company out of debt and upon a safe footing. But the abuse of the assessment privilege by dishonest officers, in calling assessments when none were needed, for the purpose of manipulating the price of shares or depressing the stock market; calling one assessment after another, until the patience and pockets of small shareholders were exhausted, forcing them to sell, has been such a common trick of the operators on the Pacific Coast that the assessment plan has been brought into disrepute. The laws of the State of New York and other Eastern States, and those of Colorado, prohibit the assessment of the shares of corporations; and, of late, most mining companies are organized under the non-assessable plan. The mines on the Comstock Lode were for a long time, and are still, in a great measure, the favorite field of operations for the sharpers of the Pacific Coast.

The following sketch of the history of two mines on the lode will give an idea of the great fluctuation in prices of shares, largely owing to the manipulation of stock operators and the officers of the mines. The Consolidated Virginia Mining Company was incorporated June 7th, 1867, owning five claims, aggregating one thousand one hundred and sixty feet, on the Comstock Lode. At that time each foot on the lode was represented by one share of stock, par value one hundred dollars. On July 6th, 1868, the management increased the number of shares to ten shares for each foot, adding two hundred shares beside, making a total of eleven thousand eight hundred shares. This number was doubled four years later, making twenty-three thousand six hundred shares, two thousand of which were deducted for owners of claim known as "Central No. 2," who refused to accept the new issue. This left the Consolidated Virginia with twenty-one thousand six hundred shares, and one thousand and ten feet on the lode. The third increase was made September 16th, 1873, when the twenty-one thousand six hundred shares were multiplied by five, making one hundred and eight thousand shares. The fourth increase was made in March, 1876, when it was again multiplied by five, making the capital stock five hundred and forty thousand shares, of the par value of one hundred dollars. In other words, the capital had been increased proportionately from one hundred and one thousand to fifty-four million dollars. But, in the meantime, the stock had been still more inflated by the organization of the California Mining Company, on the same lode, the Consolidated Virginia parting with three hundred feet of its property, which was put into the California property, which was stocked also at five hundred and forty thousand shares, with only six hundred feet on the lode. The following tables show the monthly fluctuations in these shares for the years 1874 and 1875, and the consequent rise and depreciation in values:

1874.	PER SHARE. Highest.	Lowest.	1874.	PER SHARE. Highest.	Lowest.
January	$97 50	$4 00	July	$85 00	$75 00
February	70 00	62 50	August	84 00	73 00
March	86 37½	67 50	September	94 00	80 25
April	90 00	82 50	October	120 00	91 00
May (1st div.)	85 00	66 50	November	188 00	121 00
June	84 37½	78 00	December	590 00	185 00

The fluctuations in Consolidated Virginia and California in 1875 were as annexed:

Month.	CON. VIRGINIA. Highest.	Lowest.	CALIFORNIA. Highest.	Lowest.
January	$700	$500	$780	$300
February	518	360	*80	46
March	470	405	68	52
April	460	440	66	61
May	460	395	64	63
June	422	300	61	53
July	341	297	65	55
August	360	240	78	55
September	280	245	60	51
October	335	210	65	51
November	400	237	76	50
December	511	360	77	65

* New stock—5 shares for 1 share.

Since, on a basis of 540,000 shares:

	CON. VIRGINIA.		CALIFORNIA.	
	Highest.	Lowest.	Highest.	Lowest.
1876	$96 50	$35 75	$94 00	$44 00
1877	55 00	21 50	53 00	......
1878	26 00	8 87½	34 00	12 00
1879	9 87½	4 50	11 50	4 50

In the year 1879, one hundred and thirty-eight mines of the State of Nevada levied assessments, amounting to eleven million four hundred and four thousand four hundred dollars;* the larger portion of this amount was levied upon the Comstock Lode. The total dividends from the incorporated dividend-paying mines of the United States for 1879 (not including profits from private companies), was eleven million ninety-four thousand four hundred and twelve dollars,† or about three hundred thousand dollars less than the assessments upon the mines of a single State. Yet the State of Nevada produced in gold and silver bullion, in 1879, almost twenty-two millions,‡ or about double the amount of assessments. Why this lack of dividends, unless it be swallowed up by dishonest management or exorbitant salaries to mining officials? But, although the assessment laws of the Pacific Coast have been a great aid to the sharpers in "freezing out" small stockholders,

---

* San Francisco Stock Report, December 22d, 1879.

† Mining Record, New York.

‡ Twenty-one million nine hundred and ninety-seven thousand seven hundred and fourteen dollars. Mr. Valentine's Report of Wells, Fargo & Co.'s Express, S. F.

by forcing the market down, and compelling weak shareholders to sell by burdening them with assessments, yet sharp practices are not confined to assessable stocks, nor to the Pacific Coast.

We give the case of the Little Pittsburg management, as one instance among many which might be cited of tricky management with a really valuable mine, of which the stock is unassessable. The Little Pittsburg Consolidated Mining Company, of Leadville, Colorado, was organized with a capital of twenty millions, divided into two hundred thousand shares, of the par value of one hundred dollars each. After paying thirteen regular monthly dividends, aggregating one million two hundred and fifty thousand dollars, up to the middle of January, 1880, when the stock was selling for thirty dollars per share, making the current value of the mine six millions of dollars, the stock suddenly declined. The mine continued its regular dividend of one hundred thousand dollars for February, and its average daily output of nearly one hundred tons of ore; yet, with no apparent reason, the stock, February 20th, had declined to twenty-one dollars per share, and a week later to fifteen dollars; or, in a month, the mine had depreciated in value one-half. By March the 1st it had fallen to twelve dollars and a half, and operators wondered at the big decline, while the production continued to be large. It then began to be rumored that the ore reserves were giving out, and by the middle of March it was as low as eight dollars. And the directors of the company suddenly stopped dividends. Everybody wondered. The papers heretofore had been full of the vast hidden wealth and ore reserves of the Little Pittsburg. The loss to stockholders was immense. The stock continued to decline, until in May it was down to about five dollars. In the meantime it had been discovered that the heaviest stockholders, who were directors and officers, had disposed of eighty-five thousand shares of stock between February 1st and March 13th, previous to the great decline, and that they were now among the smallest stockholders, having been quietly disposing of their shares, which being thrown upon the market, had been the prime cause of the great decline. Another fact was also developed, by the examination of an expert of the underground workings of the mines, that the superintendent, in order to pay the large dividends, had been allowed to exhaust the amount of ore in sight in the mine, and through mismanagement had not properly explored and opened up new ore-bodies ahead of the production, although there were undoubtedly rich ore-bodies remaining. In other words, *the mine had been "managed"* to suit the ideas of the heavy stockholders, who wished to dispose of their stock regardless of the future good of the mine. Had the last few dividends been smaller, and the surplus money used in furthering development and exposing new ore-bodies, so that dividends need not have ceased altogether, this sudden loss and depreciation to the stockholders need not have occurred. Above all, had the mine, instead of being capitalized at the greatly-inflated sum of twenty millions, been put at *one-tenth* of that amount, the advantages over the plan which was adopted would have been incalculable.

A Very Common Swindle

is from what are called "Wild-Cat Mines." An advertisement appears in all the mining journals, and mercantile newspapers and others, giving a glowing account of the rich assays from the mine or mines in question, describing it as in close proximity to some well-known, valuable mine, or as upon some "great mineral belt," or as an extension of some valuable lode. A prospectus is issued and sent out, giving the graphic description of some professed mining engineer and expert, who has examined the property, and who, although his name is unknown to the public, and far from familiar, finds "millions in sight" in the lode. The capital stock is one million dollars, divided into one hundred thousand shares of ten dollars each (which is a very common way of capitalizing mines). "It is expected," the circular reads, "that the stock will speedily advance to par as soon as the mill is completed and begins to crush ore; but in order to build a mill and further develop the mine, twenty-five thousand shares will be disposed of at the low price of two dollars per share; and when these are disposed of, the price will be advanced to three dollars per share, when twenty-five thousand shares more will be offered." The innocent investor pays his money and receives some finely-lithographed certificates of stock, duly signed by the officers of the company. A few months pass away, and nothing is heard from the mine. Finally the investor writes to the secretary of the company for information, and is told that "sufficient stock not having been sold to erect a mill," work is not progressing at the mine. "But as soon as the mill is erected," dividends will soon be forthcoming. A few months more elapse, in which nothing is heard from the mine, when the investor writes to some reliable mining journal, and is told that the mine is a "Wild-Cat;" that there is such a hole in the ground in some mining region; but there is nothing in it, and that the officers are unknown and irresponsible persons; that his stock is worthless; the officers having pocketed the money. Investors should buy no shares in mines not well known, of whom the officers are strangers, without the advice of professional mining men or experts, whom *they do know to be reliable*, and who can recommend the property. Even then it may be difficult to always avoid wild cats.

Tunnel Schemes.

Another species of "feline" is in the well-advertised shares of tunnels to pierce a "well-known mineral belt." A tunnel is started in some mining region, in which the mines are well known to the public; such as the Black Hills, Nevada, or the San Juan region, Colorado; and it is stocked and the shares offered for sale. The tunnel is advertised as *sure to pierce a mineral vein* "well known to exist as a continuation" of some valuable lode, at a certain distance in, and accordingly a few thousand shares are disposed of, the money expended in driving the tunnel and in paying the salaries of officers. No mineral is found, and the stock becomes worthless.

Moral: There is no safety in investing in a tunnel or mine *in which there exists no paying mineral when the investment is made.* Mining stocks have been advertised and sold, of which no such mine as represented could afterward be found, and no *bona-fide* officers of the names given on the certificates. It is a very common scheme to buy up cheap prospect-holes, in which there is no mineral of consequence, take them East and stock them at a million, and offer the stock upon the market. Hundreds of mines are stocked in New York, which, it has been reported, cost the promoters one or two hundred thousand each, for which they actually paid but from ten to twenty thousand for, and which were stocked at a million, and the stock disposed of at a round price. The profits to such schemers, when successful, are immense. For instance, a mine purchased for twenty-five thousand dollars is stocked with one hundred thousand shares, at the par value of ten dollars each, or a capital stock of one million. The mine may possibly prove good.

Twenty-five thousand shares are offered at two dollars per share, and are all taken, ...	$50,000
Twenty-five thousand shares more are offered at five dollars, and sold (which is not an unusual case when the mine shows up ordinarily good), . .	125,000
Total receipts for half the mine, .	$175,000
Deduct cost, ...	25,000
	$150,000

Profits from sale of half the mine are one hundred and fifty thousand, and the promoters have still half the stock left. This is but a fair showing of the manner in which many mines are stocked, and it will be seen that the investor in stocks has to pay frequently *many times what the mine cost*, or is really worth, *and always pays the promoter too high a profit* on his investment.

A very common way of deceiving stockholders

into purchasing stocks at greatly inflated prices is through hired or bribed experts or engineers, who, for a large consideration, are procured to examine and report upon the property, and who represent to the public that there are "millions in sight" in the mine; giving to the property a fictitious value, and a consequent inflated price to the stock.

There seems to be no manner in which the "insiders" or the "management," who are the officers of mining companies, can be prevented from manipulating or controlling the prices of mining stocks, if they are so disposed. The profits from investments in stocks are, therefore, very uncertain, and while there are many honestly, well-managed, and well-officered mines, which offer as safe investment as any other business, there seems to be no rule to give the inexperienced whereby they may invest with safety, except to keep out altogether, unless they seek a controlling interest or have trusted friends who control the corporations in which they invest. To the experienced operator in stocks we need give no advice. It is his profession, into which he has been educated by long experience, and he can readily judge for himself.

With the explanation of one more trick of sharp mining men we conclude this chapter. In a few words, it is salting a mine with good ore, or silver filings, to effect a sale of the property. We quote from the San Francisco *Stock Report* a graphic description of a salted mine on the Comstock Lode, in the olden time :

Salting a Mine.

"We hear of wild-cats now, but 1863-64 was the time for wild-cat mines. Most of the so-called wild-cats of to-day are prospecting propositions, honestly striving to develop a paying property, and having location or indications to justify an expenditure of stockholders' money in seeking for bonanzas. In those days a wild-cat was a clear case of "salt," and thousands of confiding people parted with their money for shares in salted claims boomed up to high figures on the strength of high assays. There was one claim, called the North Ophir. It was north of the Ophir, it is true, but a long way north; and the mere fact that, being so far removed from the famous Ophir mine, it was called North Ophir, was in itself a suspicious circumstance. As a mine the North Ophir was not worth three cents a mile, but as a speculation it was a success—to the insiders. Some bad, unscrupulous man stole into the North Ophir shaft one night, and poured a lot of melted silver and silver filings liberally into the seams and cracks in the ledge, where it was exposed in the bottom of the shaft, and rubbed it into the rock at the bottom and sides. A few days afterward a party of experts were invited to visit the mine. They went to it, entered the shaft, examined the ledge, which was in reality as barren as a mule, and took samples of rock for assay. The assays panned out big; the experts posted their friends; there was a rush for the stock, an excitement, and in a few hours North Ophir bounded from three to thirty dollars a share, and hard to get at that. There promised to be a North Ophir boom, and mining people were talking of North Ophir mounting away up into the thousands. But an unfortunate circumstance (for the North Ophir people) pricked the bubble. One of the experts, who had taken samples from the shaft, examined one of the pieces of the ore. Behold, there was the tip of the wing of the American eagle, as represented on the American half-dollar, impressed on the rock. A closer examination gave convincing proof that the rock was salted. The expert did not keep the news to himself. The science of shorting mining stock had not then been developed—though we believe Jack McKenty, or some equally cute individual, did sell a few shares, "to arrive"—and North Ophir, which had gone up like a rocket, came down like a stick as soon as the news got out that the mine was salted. The stock sold in the morning for thirty dollars a share; that night you could buy the stock in big blocks at two dollars a cord."

CHAPTER XXII.

HISTORY OF A FEW MINING MILLIONAIRES; MEN WHO MADE THEIR FORTUNES IN MINES—LIFE OF LIEUTENANT-GOVERNOR H. A. W. TABOR, OF COLORADO; EX-UNITED-STATES-SENATOR JEROME B. CHAFFEE, OF COLORADO; GEORGE H. FRYER, ESQ., OF COLORADO; ARCHIE BORLAND, ESQ., THE OWNER OF SOME BLACK HILLS BONANZAS; JOHN W. MACKAY, OF CALIFORNIA; JAMES C. FLOOD, ESQ., OF CALIFORNIA; JAMES G. FAIR, OF NEVADA; UNITED STATES SENATOR WILLIAM SHARON, OF CALIFORNIA—SANDY BOWERS, A COMSTOCK CHARACTER SKETCH.

Governor H. A. W. Tabor.

HON. H. A. W. TABOR, Lieutenant-Governor of Colorado, possesses one of the most suddenly-acquired fortunes of any man in the United States. A little over three years ago Mr. Tabor was the proprietor of a small grocery store at Fairplay, Colorado, and being unostentatious and closely attentive to business, remaining at his desk and always at home, he was not known outside of his circle of business acquaintances. Little is known of his early history, or from whence he came, until we find him in California Gulch, in the little town of Oro, in 1860, where he lived with his family and kept a small store for several years. It is said that his wife was the first woman who entered California Gulch. After the placers of the gulch began to be worked out, and the population began to scatter, we believe he removed to Fairplay, where he was engaged in business until 1877, not, however, being blessed with more than moderate means at this time. The first intimation of carbonate discoveries in California Gulch drew him to Leadville, and in July, 1877, he established the second store erected on the site of Leadville. In May, 1878, August Rische and George T. Hook began to sink a shaft north of Stray Horse Gulch, near Leadville, at a place since known as Fryer Hill. They had no money to obtain supplies or tools for prospecting, and Mr. Tabor furnished them from his store what they needed, they agreeing to give him a one-third interest in whatever they might discover. It cost him about seventeen dollars to outfit the party to commence work.

It happened that the ore strata was unusually near the surface at that point, and that Rische and Hook reached it in a very few days. The first wagon load of ore netted the owners two or three hundred dollars, and as they drifted on the ore strata it grew larger, until it was several feet thick, and the owners had "struck" a fortune. The location was called the Little Pittsburg. A force of men was put on the mine in June, and it was soon yielding seventy-five tons of ore per week, which milled from ninety to over two hundred ounces of silver per ton. The Little Pittsburg was, therefore, paying thousands of dollars per week in profits in August and September. In September

Tabor and Rische bought out the one-third interest of Hook, paying him ninety-eight thousand dollars in cash, mostly from profits which they had already received from the mine. The buyers cleared the purchase-money in three weeks thereafter. By the first of November the receipts from the sale of ore had reached the sum total of three hundred and seventy-five thousand dollars, all in a few short months, and the fame of the Little Pittsburg was about to make Tabor lieutenant-governor of the State. Everything that Mr. Tabor turned his hand to thereafter seemed to prosper. Beginning with his investment in Little Pittsburg, he kept making investments in other mines and real estate in Leadville, and kept adding to his wealth until he is the richest man in the State. There is no telling what he is worth; the figure can be placed at ten millions, and then it will not be too high. All this wealth acquired within the short period of three years! His income from the numerous mines in which he is interested at Leadville and elsewhere cannot be less than five thousand dollars per day, and the mines are constantly growing more productive and their value increasing every day. Mr. Tabor sold his interest in the Little Pittsburg property, realizing therefrom something like one million three hundred thousand dollars. But before parting with it he had taken out hundreds of thousands of dollars. He has purchased property all over the State of Colorado, and it is said was offered a million and a half for his interest in the Chrysolite, at Leadville. His mines in the San Juan country are reported to be developing very rich.

Mr. Tabor divides his time about equally between Denver, where he lives and owns a splendid residence, and Leadville and the eastern cities. He is now owner of a great amount of property and real estate in Denver and other places. It is said that he recently purchased the whole of South Chicago, paying a round million for it. He is the largest holder, with one exception, in the First National Bank of Denver, and owns large quantities of real estate in that city. The Tabor Block, opposite the First National Bank of Denver, is one of the finest business buildings in the West. In Leadville Mr. Tabor owns much valuable property outside of the mines. The New Opera House, in that city, was put up by him, and it is a structure which would do credit to any city. Mr. Tabor has proved to be a public-spirited citizen, and to be generous and ever disposed to relieve the distress of the poor. Personally, Mr. Tabor is a rare good fellow. Never did Dame Fortune bestow her favors upon one who accepts them with less ostentation and parade than does Mr. Tabor. He is popular everywhere, and his wealth has only added to his worth, in that it has given the public a greater opportunity than heretofore to know and appreciate him.

Jerome B. Chaffee.

Senator Jerome B. Chaffee, of Colorado, was born in Niagara County, New York, April 17th, 1825. He had an academic education, and when quite young removed West, locating in Michigan, and afterward removing to St. Joseph, Missouri, where he was engaged in the business of banking. In 1857 he organized the Elmwood Town Company, in Kansas, and became its secretary and manager. He went to Colorado in the spring of 1860, and went at once to what is now Gilpin County, and entered upon the work of developing some prominent gold lodes which he had secured. In company with Mr. Ebin Smith, a skillful mining expert and manager, he erected the Smith & Chaffee stamp-mill. This enterprise proving successful, did much to revive the drooping courage of the miners in the vicinity, and gave an impetus to the mining industries of Gilpin County.

In 1863 he sold his interest in the lode he was then working, but subsequently re-purchased it, and consolidated it with other mines, the whole constituting what has since been the famous Bobtail Mine and Tunnel; its name being derived from the fact that a bobtailed ox, harnessed to a drag, consisting of a raw-hide stretched across a forked stick, was used for hauling the pay-dirt from the mine to the gulch for sluicing.

In 1869, a consolidation of various interests on the lode was effected by Mr. Chaffee, who became the heaviest stockholder in the Bobtail company, which was the best known and most prosperous mining company for a time in Colorado, producing annually from three hundred thousand to five hundred thousand dollars in gold. In 1876, Mr. Chaffee was elected a United States Senator by the first Legislature which convened after Colorado was admitted as a State.

In the summer of 1878, Mr. Chaffee went to Leadville, and made some investments there. George H. Fryer, in company with a man named Borden, owned the "New Discovery" mine, on Fryer Hill. Senator Chaffee bought Fryer's interest for fifty thousand dollars. Before Chaffee left the camp, H. A. W. Tabor and August Rische, who then owned the adjoining claim—Little Pittsburg—fearing that the older title of the "New Discovery" claim might imperil their claim, took it off his hands at one hundred and twenty-five thousand dollars, thereby giving him a profit of seventy-five thousand dollars on a few days' investment. Subsequently Chaffee, in company with others, bought August Rische's entire interest in the "Little Pittsburg" and "New Discovery" mines for two hundred and sixty-two thousand five hundred dollars, and thus became half owners in these claims with Tabor. On the 18th of November these parties organized the Little Pittsburg Consolidated Mining Company, which was a consolidation of the New Discovery, Little Pittsburg,

Dives and Winnemuc. In the next five months the actual yield of ore from the consolidated mines, at the sold or assay value, was nearly a million of dollars, and the profits in the great rise of stocks to the promoters were enormous. Senator Chaffee probably realized, as his share of the profits, not less than two millions from this transaction.

Senator Chaffee has probably made larger investments in mining operations than any other man in Colorado. He owns, it is said, about a hundred gold and silver lodes, in various stages of development, among which is the well-known Carribou silver mine, in Boulder County, besides other interests in Leadville and in the Little Pittsburg.

George H. Fryer.

George H. Fryer, from whom the now famous Fryer Hill of Leadville takes its name, was born in Philadelphia, September 4th, 1836. He received his education in that city, graduating at the public High School; after which he accepted a clerkship in a Philadelphia silk house for about four years. In 1857 he started West, first locating at Leavenworth, Kansas, where he was engaged in the real estate business up to 1861. In the spring of that year he went to Denver, and soon after went to Summit County, and during the following summer engaged in placer mining with moderate success. The following autumn he went to the Montgomery mining district, where he was elected mining recorder, and continued to discharge the duties of that office until the winter of 1862, when the Legislature abolished the office.

Mr. Fryer was engaged in mining up to 1864, when he returned to Philadelphia and made a sale of some mining property, realizing therefrom a small fortune. He remained in the East but a short time, but returned to Colorado, and has since been actively engaged in mining. On the 4th of April, 1878, he struck ore in the "New Discovery," now a part of the Little Pittsburg Consolidation, at Leadville. The hill took the name of Fryer Hill, and as long as the mountain remains it is probable that the name of George H. Fryer will not be forgotten. Fryer soon sold his interest in this wonderful mine to Chaffee for fifty thousand dollars, and soon afterward bought back one thousand shares of the stock of the Consolidated Little Pittsburg Company, at the rate of something like four million dollars for his mine, and got a good bargain then, from which he subsequently received large profits.

Mr. Fryer is now very largely engaged in mining, having interests in a large number of Leadville bonanzas, from which he receives a daily income amounting to a small fortune. But he is the same man to-day that he was before striking his bonanzas. He is not purse-proud, and he gives liberally of his means to all public and charitable institutions. He is, indeed, the poor man's friend. Mr. Fryer is highly esteemed as a citizen, having hosts of warm friends, and is popularly spoken of by his party (Democratic) as a candidate for Congress at the next election, which is indicative of the high regard with which he is held. He lives in the city of Denver.

Archie Borland, the Owner of some Black Hills Bonanzas—His Early Career and Good Luck.

There are few men on the Pacific Slope better known in mining circles than Archie Borland. Mr. Borland came from Ireland a mere boy, with all his capital invested in good muscle and a clear head. He settled down on a farm in a small place a few miles below Albany, New York, and worked away until the mining fever swept over the land in 1849. The epidemic soon made its way up the Hudson, and he was among its first victims. He heard of the wealth that was being picked up loose all over the soil of California, and desired to do some of the picking himself. But it was a long journey to California—much longer than now, and took a great deal of money to get there. Not prepared to start at once, he immediately began to save up his earnings, with the hope of gathering enough in a short time to pay his passage-money. Ten dollars a month was good wages on a farm in those days, and men seldom received over twelve, so that saving was slow work. It was not till early in 1852 that he was ready to start, and he took passage in one of the steamships that were running to California, and landed without accident in San Francisco.

"I made up my mind to two things at the start," said he; "not to work for wages any longer than I could help, but to go in for myself; and then, never to take a partner." This plan served him well.

With nothing but a frying-pan, tincup, pick and necessary tools, with a small supply of provisions, he went, upon reaching California, into Nevada County, and began work in the mines at four dollars per day. Working there, often in water up to his waist, at the hardest kind of labor, with the plainest of food of his own cooking, he carefully laid by the greater part of his wages, intending some day to own the mine he was working. But the mines did not prove profitable enough, and as long as he received his four dollars a day he kept to work, not deeming it a profitable investment. After some months' labor he went to Grass Valley district. Here he was able, in a short time, to buy, in a small way, into some claims, and made some money.

For several years he continued at this, growing a little richer every year till 1858, when the Frazier River excitement began. Parties of miners were going there every day, and wild rumors of their tremendous success came back. The ground was believed to be covered with gold, and Mr.

Borland caught the fever with the rest, and sold out in Grass Valley and went. The steamer on which he took passage made her way several hundred miles up the river, and the men got ready to pick up a little of the loose gold. But there was little or none to pick up. The expedition proved a disastrous failure, and the young miner soon found that he had sold out a good thing to take his chances in a sinking ship.

Reaching Grass Valley again in 1859, with just money enough left to buy a small interest in the old mines, he set about retrieving his fortune. One of his speculations, after getting a little on his feet again, was the buying of the Rock Tunnel, which had been opened two thousand eight hundred feet. In 1863, some speculations went wrong; a vein was not as wide or as valuable as expected, and what money Archie Borland had saved disappeared like snow in the sun. Everything went with a crash, and after eleven years of hard work he was once more reduced to working for wages. "The great thing," said he, "was never to quit. Stick to it. Never let go your hold, and you are bound to catch something."

With precious few dollars left to operate with, he bought a mule, and once more wore a tincup in his belt and a frying-pan strapped to his blankets. With a small party of men, bound upon the same errand, he mounted his mule and started across the hills for Idaho. They had eighteen mules in the party, packed with provisions. There was not a house on the way, and he rode that mule nine hundred miles, going from one mining camp to another, but always in the direction of Idaho. After a long and painful journey he reached Idaho, and went to work. By diligent labor he soon owned shares in some of the mines there. By 1866 he had accumulated fifteen thousand dollars, which he had not in checks, or drafts, or bits of paper, but in gold dust, done up in bags. With his stock of gold dust, Mr. Borland determined to return to California, which he considered his proper field, although he had prospered in Idaho. Wells, Fargo & Co. had then an established express line into Idaho, and he went to them to learn what it would cost to transport his dust to San Francisco. He was told one dollar and twenty-five cents for every one hundred dollars' worth, and even at that price he must run his own risk. He thought, however, that if he had to run his own risk he might as well save the express charges, and transport it himself, for he knew that in such a case nobody would get his gold without fighting for it. The road at that time was infested with "road agents," and the journey was a dangerous one. But he bought some mules, a big navy revolver and a supply of provisions, and started. It was a long and lonely journey. But after six nights of hard riding he landed his gold dust safely in San Francisco. Then he bought into Comstock and other mines, and made money rapidly. There were no mining stocks in those days, but a "foot" of surface in length on the vein was the equivalent of a share of stock. One of his first successful speculations on a large scale was in the Savage mine. In July, 1866, he bought five feet of the Savage, at one thousand one hundred dollars per foot. During the first month that he owned it he drew fifty dollars per foot dividend. The second month the dividend rose to seventy-five dollars; then it took another rise, and reached one hundred dollars a foot dividend every month. In the following year, 1867, the value went up tremendously. Early in the year it was worth two thousand dollars a foot, and from this it rose steadily to two thousand five hundred, three thousand, four thousand, and five thousand dollars a foot. At this price he sold his five feet for twenty-five thousand dollars; having received more than his original investment in monthly dividends before selling. Other stocks that he had went up, and his fortune gradually increased, until from twenty-five thousand dollars it went gradually to fifty thousand, seventy-five thousand, and four hundred thousand dollars. In 1869, he bought largely in the Comstock mines, and up to 1872 he had no competitor in his gigantic operations. Then Jones and Heyward came up, and, as he says, he had to keep his eyes open from that time on, "for I had to compete against men with millions," he said, "and who knew how to use them. I went into Crown Point shares, and got singed. But that was nothing; there's ups and downs in any business." At one time his dividends from California and Consolidated Virginia amounted to forty thousand dollars a month.

"The best winning speculations I ever made," he said, "were from 1872 to 1874. I made a heap of money in that time, and I needed it to carry me through my worst losses in 1878. But I got through, and am still afloat. In 1872 I bought five hundred shares of 'Central' for less than five thousand dollars. This mine was soon afterward cut up and consolidated with California, and the redistribution gave me one thousand two hundred and fifty shares. I never paid an assessment on it, for none were called, and in a short time the stock went up to seven hundred and forty dollars a share, when I sold. That paid pretty well. I paid five thousand dollars, and sold for nine hun-hundred and twenty-five thousand dollars. That left me a profit of nine hundred and twenty thousand dollars.

"That was one of my best speculations, but not quite as good as another venture made about the same time. I bought five hundred shares of Consolidated Virginia for fifty dollars a share, and paid two assessments, which brought its cost up to fifty-six dollars per share. This mine was also 'cut up' about this time, and my five hundred shares spread out into two thousand seven hundred and fifty shares. I held this stock for two years, when it went up like a flash. I sold out in 1874 for seven hundred and fifty dollars a share. It cost me twenty-eight thousand dollars for my stock, and I sold it for two million twenty-six thousand five hundred dollars. One Saturday the stocks were selling for three hundred and fifty dollars per share; but I held on. The following Monday it was five hundred dollars a share. I tell you, it took a pretty strong head to stand that. I was getting rich, not at the rate of a million a minute, as they say, but one hundred thousand a day or more; but that was fast enough. Nobody ever knew anything about this till it was all over. I never told my wife and family about my business affairs, and never kept any books. The only book that I kept was a little pass-book, that I carried in my inside coat pocket, with a record of what I bought and sold, and the price. I never let any one see this, and never brought it out, except on a Sunday afternoon or when I was alone. My heavy losses in 1878 were on Sierra Nevada and Union. I lost one million four hundred thousand dollars, then, inside of ten days; shrunk that much in depreciation of stock."

Mr. Borland is still interested in a good many mines, though he spends more time in his handsome residence, in Oakland, California, with his family, than when he was younger.

John W. Mackay,

the subject of this sketch, was the youngest of the Mackay-Flood-Fair combination, who were long kings of the Comstock Lode. He was born November 28th, 1835, in Dublin, Ireland. He came to America a mere lad, and for some years found employment in the office of William H. Webb, the famous shipbuilder of New York. In the fall of 1852, he joined a party bound for California, and about the close of that year he reached his destination, having made the passage in a vessel built by his former employer. Young Mackay immediately engaged in placer-mining at Alleghany, Sierra County, California, where moderate success rewarded his efforts. He did not drift into mining, and delve among the rocks and sand, as a make-shift, as was the case with a majority of those who arrived there in the golden age. He entered mining as a profession; he entered it to stay, and he did stay. He did not have luck above that of other men; but, on the contrary, had in his early mining career his full share of misfortune. He did not lose a fortune in those early days, for the best of reasons: he had none to lose. But this fact did not prevent him from undergoing the many privations incidental to a miner's life in California and Nevada. Fortune smiled slightly on Mr. Mackay while at work in the placers of Sierra County, and although it was a sickly sort of a smile, still he managed to secure sufficient money to proceed to Virginia City, Nevada, and inaugurate an enterprise of his own.

In connection with a partner, he started a tunnel in what was then known as the Union ground, north of the Ophir mine. Here fortune frowned on him. His funds were speedily exhausted, as a matter of course, and once more he started at the foot of the ladder for a competency. He made his brawny arms earn him four dollars a day as a timber-man, in the Mexican mine. He also swung a pick and shovel in some of the mines at the same wages. His ideas of wealth were very moderate in those times. There was no limit to his ambition, if we are to accept the narratives of his old comrades; but his chief desire was to accumulate twenty-five thousand dollars, for the purpose of assisting his beloved mother in her declining years. This desire affords a good index of the man's character. He was just the kind of a man to express such a desire. His fidelity and rugged integrity are as prominent to-day as they were when he worked for wages and earned all that he received. The first substantial start made by Mr. Mackay was in connection with the "Kentuck" mine, in Gold Hill. After many changes of fortune, he became interested with Mr. J. M. Walker, in 1863. This firm was enlarged by the addition of Messrs. Flood & O'Brien, in 1864, and so continued until 1868, when Mr. Fair took the place of Mr. Walker. The first few hundred thousand of the now stupendous wealth controlled by this firm was made during their control of the Hale & Norcross mine, from 1865 to 1867. The career of Mr. Mackay, since that time to the present, has been chiefly noticeable for the active part he has taken in the incessant and successful efforts made by the partners to obtain further acquisition of territory on the Comstock Lode. Backed by their constantly-increasing capital, their efforts have resulted in opening to the world the wonderful "Consolidated Virginia" and "California," known as the Bonanza Mines. On the 25th of November, 1867, Mr. Mackay married an accomplished lady, the daughter of Colonel Daniel E. Hungerford, of the army. They have two children. Mr. Mackay's home is in Virginia City, and during most of the time since the discovery of the bonanza in the Consolidated Virginia and California he has largely directed the affairs of the firm, in which he had a three-eighths interest. His decision of character, his shrewdness, his application, and his personal popularity, have largely contributed to the renowned success of what is known as the bonanza combination. He is one of the few men whom fortune has not spoiled. He has stood the test well. His liberality has again and again placed old friends in affluent positions. We could illustrate this generous nature by numerous statements of facts which would surprise the world; but, in deference to Mr. Mackay's feelings, and with due regard for the proprieties, we abstain from giving details.

Whether at his mansion in Paris, or in the dripping depths of the mines, he is the same quiet, manly, unostentatious John Mackay. For some time Mrs. Mackay has resided at the French capital, where the children are being educated, and where, it will be remembered, she entertained the late Chief Magistrate of the United States, General Grant, in princely style.

We cannot dismiss the subject of this sketch without saying a word as to the manner of the discovery of the great bonanza. It is a common idea that the bonanza discovery was simply a piece of luck; such luck as might befall any miner. The circumstances that preceded this discovery are overshadowed by the importance of the find itself. Few people realize the difficulties attending the search for this marvelous body of ore. When Mr. Mackay and his associates inaugurated their search for the bonanza, the ground involved in their prospecting operations had been abandoned by Sharon and other large operators. Mr. Mackay did not begin work with the definite knowledge that he was bound for a bonanza, but believing in the richness of what is now the bonanza territory, Mr. Mackay and his associates purchased the controlling interest in the corporation then owning the ground, and began to hunt for the ore body. The tax upon the patience, endurance and skill of the firm, as well as upon their purses, can barely be imagined. They paid out not less than five hundred thousand dollars to carry on this prospecting operation. Prominent men in the district, who had succumbed to discouragements, looked with disfavor on the work being done by Mr. Mackay and his partners. It would not pay the trouble and expense, they said. But Mr. Mackay continued his work. That famous drift from the "Gould and Curry" mine, through "Best and Belcher" ground, into the Virginia, was run over one thousand two hundred feet before it made the strike that gave the bonanza to the world. The firm were suddenly enriched by millions. What trials and tribulations, what heart-aches, hopes and fears were involved in that long drift we may never know. That the work was under the best of management, however, has always been an acknowledged fact. The results of that management have made an imperishable impression, and to John W. Mackay that impression owes its existence: he led the forlorn hope. In closing our sketch of Mr. Mackay, we may remark that he is of herculean build, easy of carriage, and has a genial presence. He looks the man he is—the prince of miners and boss of the big bonanza.

James C. Flood,

another mining prince, was born in New York City, about 1826, and is now in his very prime. His early education was most practical in its character. He did not enjoy the benefits of a collegiate course, but the instruction he received in the ordinary English branches was systematic and thorough. In 1849, he took passage for California in the ship "Elizabeth Ellen." His brains constituted his sole capital. In 1854, he associated with the late W. S. O'Brien, and the famed firm of Flood & O'Brien was formed. The first notable enterprise engaged in by the firm consisted of operations in the "Kentuck," and other mines on the Comstock, in which they generally contrived to secure a controlling interest. This was as early as 1862. Their operations in the "Hale & Norcross" mine, a few years later, was on a scale so large as to attract general attention to them as mining speculators. The operation which finally made the name of this mining firm known throughout the world, was compressed within the short space of a few months in the early part of 1875. The existence of those vast bodies of ore in the "Consolidated Virginia" and "California" mines, on the Comstock Lode, which Mr. Mackay also assisted to develop, was suspected as early as February, 1874, and were made certain by the proprietors in December of that year. The generosity with which they dealt with those who had the good fortune to be their friends, was very generally acknowledged. They were not content to see their fortunes growing with colossal strides each hour, but desired all who had been kind to them in the past to accompany them on the road to prosperity. The establishment of the Nevada Bank, in San Francisco, was the idea of Mr. Flood, and was organized by the firm.

In appearance, Mr. Flood is prepossessing, strong, and about five feet ten inches high, with robust form. Mr. Flood became the leading financier of the Pacific Coast. His power of mental concentration, his quickness of perception and his liking for finance, combined to make him the financier of the Bonanza firm. He is a persistent and intelligent student of financial history, and keeps himself thoroughly informed on the progress of finance. Mr. Flood is an incessant worker during business hours. He is, however, devotedly attached to the home circle, and his surroundings are of the most pleasant character. He spends the winter months at his plain and unostentatious residence in San Francisco, but the summer nights are passed at San Mateo, where Mr. Flood has one of the most attractive country seats on the Pacific Coast. His family consists of a wife, son and daughter, the son being about twenty-two.

James G. Fair,

is also widely known as one of the chiefs of the Bonanza firm. He was born December 3d, 1831, in Clougher, Tyrone County, Ireland, and came to this country in 1843. He attended school in Geneva, Illinois, for several years, and subsequently secured a good business education in Chicago. In common with other adventurous spirits, Mr. Fair was affected by the gold fever in 1849, and in August of that year he arrived at Long's Bar, Feather River, California. He mined on the bar for awhile, but failed to strike a profitable placer-mine. It was natural that he should turn his attention to quartz-mining. Placer-mining was conducted in a primitive style in those days, and did not afford Mr. Fair fitting opportunities for the exercise of his peculiar mechanical genius. From the placer-bar to the quartz-mine was an easy transition for him. We next hear of him engaged in quartz-mining at Angels, Calaveras County, California; and, at a later period, he figured as the superintendent of quartz mines in other counties. Even in those early days he ranked high as a professional miner. In 1855, Mr. Fair assumed the superintendency of the Ophir mine, on the Comstock Lode, and in 1857 the Hale & Norcross mine came under his direction. Among mining men he soon became accorded rank as one of the most accomplished mining engineers of America.

And he managed the mines under his charge with remarkable good judgment, and with great mechanical and engineering skill.

During their connection with the Hale and Norcross mine, the Bonanza firm, of whom Mr. Fair was now a member, secured the first half million of their princely fortune. The idea then occurred to Messrs. Flood, Fair & Mackay, of the firm, to obtain control of what was then known as the "California" and "Sides" mine, and the White, Murphy, the Central (Nos. 1 and 2), and the tract known as the Kinney ground, all on the Comstock Lode The claims were eventually secured, and to-day form the famous "Consolidated Virginia" and California mines. Previous chapters have already shown the immense and sudden wealth which afterward poured in upon the owners of these mines. The life of Mr. Fair, since this immense wealth of the great mines poured into his lap, has been that of any sagacious capitalist of large means. He began speculation in real estate in San Francisco in 1858, and owns nearly *seventy acres* in different parts of it. This property has increased in value until, of itself, it is a colossal fortune. In person, Mr. Fair is about the medium height, strongly built and, with alert carriage and pleasant face, presents an exceptionally striking appearance. He has a devoted wife and four children. His home is in Virginia City, Nevada.

Hon. William Sharon.

Senator Sharon was born in Ohio, and when a young man practiced law in the State of Illinois. Manifesting a bent toward the mercantile business, in which a brother was then engaged, rather than to the profession for which he had received his education, in 1849 he went overland to California, and at once conceived the possibilities of making a fortune as an operator in real estate in San Francisco. The records of the county since 1850 show an extent and magnitude to his transactions unequaled by any one operator in the city of San Francisco. In 1863, he was induced to take the agency and general supervision of the Bank of California, in Gold Hill and in Virginia City, in the State of Nevada.

At about the end of the first year of his agency and management of these branch banks, Mr. Ralston, the president of the Bank of California, paid him a visit. This visit happened at one of the periods of darkness on the Comstock Lode. This great silver and gold-producing belt was in eclipse. The ores of the Virginia City and Gold Hill mines had suddenly ceased to be remunerative. The mills in great number were unsupplied with ore and out of employment. The prospect seemed very discouraging and unpromising Mines of silver, which had been regarded as interminable in their endurance and extent downwards, to the consternation of all, had ceased to be remunerative or profitable to work. To the quick perceptions of President Ralston the loans in large sums upon these properties seemed lost, or nearly so, and he became apprehensive that the credit of the "home institution" would be impaired, and manifested the most decided uneasiness at the situation. The prospect of recovering the money struck Ralston on the cold side, and in his impetuous way, said to Mr. Sharon that he would feel the greatest relief if he could find a person who would assume the responsibility of returning the large sums loaned, upon a most liberal allowance of time.

Mr. Sharon at once proposed to stand responsible to the bank for the full sum, on condition that the bank advance him a considerable sum, which should be used in running certain drifts in the mines contemplated by him, and give him two years to make the total payments. Mr. Ralston preferring to look to Mr. Sharon rather than the properties acquired, and to the securities held by the bank, entered into an agreement upon the terms proposed.

Mr. Sharon here manifested something of the spirit of reliance, will-force and sagacity which subsequently distinguished him as a man of mark. He had examined, with the greatest care, every mine, in every part, and had become the mining expert as well as the banker. He had called to his aid, however, the best judgment of men experienced in mining, and made a most thorough survey of the mining properties in which he had now become interested, and finally determined to drift for a "blind ledge." A profound reasoner, he had come to the conclusion that where one man was found the country was likely to be populated; that one line of mineral gave promise of another; and that nature was not given to eccentricities in the deposit of her stores of treasure; that she might exhibit faults, but sooner or later another ore body would be discovered. A drift was at once undertaken, and prosecuted with prodigious vigor. That drift developed not only a new deposit of ore, but it produced a financial king.

The fates for once were gracious. They yielded to human will. The music of the quartz mills again resounded along the cañon. The miners were set to work again; pluck had found its reward. Other drifts were run, other deposits found, and the whole line of mines gave token of new life. Within four months from the date of Mr. Sharon's assumption of the heavy obligations he was enabled to pay the bank the whole sum due to it, and he had placed on deposit to his credit the snug sum of seven hundred and fifty thousand dollars. From this date the managers of the Bank of California came to understand that the services and co-operation of Mr. Sharon were an indispensable necessity.

Henceforth Mr. Sharon became a recognized leader in enterprises of the greatest magnitude on the Comstock, and these were managed with such adroitness and skill, that he was regarded as one of the financial chiefs of the Pacific Coast. Mr. Sharon's fame as a financier, together with his personal popularity, sent him to the United States Senate, of which he is still a member.

Sandy Bowers.

We will close with the life of Sandy Bowers, which will strongly remind the reader of the character of "Coal-Oil Johnny," of Pennsylvania, and his sudden rise to wealth and as sudden fall to poverty. It is from the San Francisco *Stock Report*, holiday number, December, 1879.

"In the earlier days of Comstock mining, Gold Hill was the attraction. The surface ores of that locality were decomposed, and therefore easily reduced, and were largely endued with gold. The claims on Gold Hill proper were small, some not over ten feet in length, comprising several series, which are now consolidated into the Imperial, Challenge, and Confidence Companies. These claims were owned mainly by individuals, by whom they were worked, and were valued at fabulous sums per foot. One of these claims was owned by an odd character, named Sandy Bowers, husband of the Mrs. Bowers who is known as the 'Washoe Seeress,' and who professes to be able, by some sort of spiritual influence, to locate the position of bonanzas, discover lost articles, read fate and peer into the future. Sandy—he is dead, now—was a rough, ignorant frontiersman, with no business ideas, and without even good common sense. But he made money hand over fist out of his mine. As his wealth grew he indulged in all sorts of nonsensical extravagance, and was bled right and left, and died the next thing to a pauper. One of his freaks was to erect a magnificent mansion of cut stone in Washoe Valley, a locality at that time all but a wilderness, which was furnished in a style of barbaric splendor—a piano in every room; hot and cold water in the parlors; its windows curtained with lace purchased by Mr. and Mrs. Bowers in Europe, and its walls hung with cheap paintings, palmed off on them in Europe at enormous prices as the originals of some of the great masters. Even a gold mine could not stand the drain put upon it by a proprietor who had no idea of the value of money, and it is no wonder that Sandy Bowers died poor. Nearly all the original locators on the Comstock either died poor or are to-day living in poverty—but none of them but Bowers ever built a big mansion. Those who did realize wealth from their locations spent it with a free hand in conviviality and generosity, after the time-honored manner of the lucky miner, who, when he makes a strike and gets his pockets full of coin, is a regular Jack ashore.

"In 1861, Sandy had accumulated so much cash that he came to the conclusion that his life would be a failure if Mrs. Bowers and himself did not make the tour of Europe. Somebody had told him that the tour of Europe was the correct thing for wealthy Americans, and he made up his mind to do it or die. Accordingly, it was announced to the small community which the population of Washoe then constituted that Mr. and Mrs. Sandy Bowers were going to Europe. Then somebody told Sandy that, in order to do the thing up in real first-class style, he should give a farewell banquet to his friends, because he might fall off the steamer and get drowned, or be shanghai'd in New York, or captured by brigands in Italy, and the banquet would keep his memory fresh in the minds of the people of Washoe. "Banquet goes," said Sandy, and banquet went. The brick International Hotel—the first brick hotel ever built in Virginia—had just been completed. It bore no more comparison to the present stately structure, which now bears its name and occupies its site, than a miner's cabin at Bodie does to the Palace Hotel in San Francisco; but it was a fine building for the early days of Washoe. In the International Hotel the banquet was held. Everybody was invited—miners and merchants, ministers and speculators, and politicians, highwaymen, gamblers, mining superintendents and other disreputable characters. The banquet was a success.

Many of the edibles were brought from San Francisco at great expense by express, and champagne flowed like the clear waters of the Carson before the march of mining enterprise had rendered them sluggish with the tailings from the quartz-mills. In response to the toast, 'Our host,' Mr. Bowers, being nudged by his nearest neighbor that he was expected to respond, arose and said:

"'I've been in this yer country amongst the fust that come here. I've had powerful good luck, and I've got money to throw at the birds. Thar ain't no chance for a gentleman to spend his coin in this country, and thar ain't nothin' much to see, so me and Mrs. Bowers is goin' to Yoorop to take in the sights. One of the great men of this country was in this region a while back. That was Horace Greeley. I saw him, and he didn't look like no great shakes. Outside of him the only great men I have seen in this country is Governor Nye and old Winnemucca. Now me and Mrs. Bowers is goin' to Yoorop, to see the queen of England and the other great men of them countries, and I hope you'll all jine in and drink Mrs. Bowers' health. Thar's plenty champagne, and money ain't no object.'"

CHAPTER XXIV.

CONCLUSION—A WORD OF ADVICE—ALL IS NOT GOLD THAT GLITTERS—AMOUNT OF FUNDS REQUIRED FOR THE JOURNEY—RAILWAY AND STAGE FARE—THE CONTINGENCIES OF SUCH A JOURNEY—SICKNESS—A NOBLE ACT—PROSPECTING A GREAT LOTTERY—WHO DRAW THE PRIZES—THE OLD-TIMERS OF '49—PLENTY OF ROOM FOR PLUCK AND ENERGY—HOW TO OUTFIT FOR PROSPECTING—WHAT TO PROVIDE—THE COST—THE BURRO—A USEFUL ANIMAL.

Conclusion—A Word of Advice.

BEFORE closing these chapters, a word of advice to the inexperienced, who, like ourselves, may have caught the "fever" to prospect for gold in the Rockies, may not be out of place. While it is all true, that we have in the Rockies, and beyond, a grand country; a country rich in minerals and gold; a country full of surprises and delight for the eyes of those who have never seen it; while there are hundreds and thousands who have sought their fortunes in this land, and have "struck it rich;" and while it is one of the most delightful excursions which Americans who are able can take for pleasure or health; yet it is equally true that "all is not gold that glitters;" that fortune does not smile on the whole multitude who woo it; that of the thousands who seek it, but a small portion wins; and that disappointment and disaster are as frequent in that land of promise as in any other.

It may be worth while, therefore, to consider well the advantages and disadvantages, "pro and con," of taking a voyage upon an unknown sea, or a journey into a new and untried land. We will, therefore, as briefly as possible, give such suggestions as occur to us which might be of service to any reader who may be contemplating a journey to the mining regions of the great West.

First, money is the all-essential element which is necessary in order to take such a trip. Without money it would be useless for young men of the Eastern or Middle States to contemplate for a moment the journey of two thousand miles or more, and the necessary preparation for the hardships of camp life in the mountains. Money, then, is the first requirement, and is indispensable to such an undertaking. Next, what amount is required? This, of course, depends upon circumstances. But for those in the Eastern or Middle States, about five hundred dollars is the sum which should be provided. It is true, that with plenty of brain and muscle, and ordinary good fortune, the trip can be made for half this sum.

The railway and stage fare to the mining regions in Colorado and the Black Hills, from either New York or Philadelphia, or other eastern cities, is about from seventy to eighty-five dollars, owing somewhat to the class of ticket purchased. From intermediate points to the same places it will be about the same from points east of Chicago. To the mining regions of Montana, Idaho and New Mexico, the railway and stage fare will probably exceed one hundred and twenty-five dollars. Therefore, when hotel bills and other necessary expenses along the way are considered, and a sum provided with which to return, two hundred and fifty dollars is hardly sufficient, though in some cases it may be enough. But there are certain *contingencies* to be provided for besides. Sickness and misfortune, and a sudden call home, should be provided for at the outset. It came within our experience to see many poor fellows sick and among strangers, without money or means of providing for their distress, and entirely dependent upon the charity of strangers or new-found friends. The condition of such is pitiable. There are no conveniences to help the sick in a new mining camp. There are no beds but the ground or hard boards, no shelter but tents and rude huts, which are no better. Tents are *very warm* while the sun beats down upon them at mid-day, hot and uncomfortable; and *cold at night*, in these regions of high elevation. The food is plain, coarse, hearty, indigestible, and unfit for the sick. While the well and strong notice none of these inconveniences, to the sick they seem many times multiplied. Very frequently medical attendance cannot be had; and when it can, costs exorbitant prices—five dollars, and often ten for a single visit. The chances are that the sick generally die under such conditions; and if they get well, the expense of a short sickness has been very great. Again, a stranger sick, and without money, will receive little attention in such a place, compared to that he would receive in many places East. Yet there are many noble acts shown by miners and others in the mountains toward those who are sick. We were cognizant of one party, a freighter, from Iowa, who had a number of teams in the mountains. Two of his acquaintances from the same place became sick, one entirely helpless, with rheumatism, and both out of money. The freighter provided money, cared for them until they were better, and as soon as they were able to start home, not having the money at hand, he borrowed one hundred dollars on the security of his teams, and sent the parties home. He remarked that he did not know as he would ever recover the money, but would not see acquaintances from his section suffer while he had means to assist them. Sickness, then, is one of the contingencies to provide for, and is our reason for putting the sum required for such a journey at what may seem too high a figure.

To those who feel within themselves the elements of success, and who, having the necessary means to go, are thirsting for adventure and burning with the fever of desire to prospect for gold in the mountains; to such we say go. The experience and knowledge of the world you will acquire, and the grand sights you will see, will more than repay you if you should not be successful in "striking it rich." But to those who would go, like many do, thinking that gold, so to speak, "grows on the bushes;" that they have only to strike a pick into the ground, or dig a shallow hole to strike a "Bonanza;" that *fortune is sure*, if they can only reach the land of promise; we say to them, *fortune is not sure;* that the search for gold is a lottery, in which the larger part of the tickets draw blanks; yet, unlike a lottery, something depends upon the brain and muscle, perseverance and pluck, of the man. It has been estimated that about one in a thousand draw prizes in this lottery of prospecting.

It can almost be said that prospecting has become a trade or profession. There are men who have followed it a lifetime, and who make it the business of their lives. We have seen those who were in the California excitement of 1849—old men now. They were in Nevada, Utah, Montana and in Colorado, in the earliest days of the Pike's Peak "boom," and later in the Black Hills and Leadville. Some of them have visited Arizona and New Mexico, and are now in the Gunnison country, still prospecting for gold and silver. A few of these own property, and have comfortable homes and families in the East and elsewhere. Yet every summer finds them in the mountains. It is a life they cannot quit, and has a power over them which they cannot resist. These are *experienced gold hunters*. They have acquired a certain practical knowledge of minerals and vein formation, which is very useful to them. *They are the first* to reach new mining regions; in fact, they are the pioneers to discover and explore the ground. It is not strange, therefore, that they possess many advantages, by their experience, over the new-comers and inexperienced "tender-feet" who visit the land. Yet many of these old-timers are poor. Some have never "struck it rich;" others have acquired a fortune, and, "Sandy Bowers"* like, have lost it. A few have still a competency. Mining regions are most always *crowded with prospectors*. There is *seldom lack of men*. There is, as a rule, enough of *these experienced ones* to thoroughly prospect the country. Here, then, is shown another of the contingencies to be considered.

It is not our desire to discourage the ambitious and plucky young man, who, with faith in his own resources, is bound for the mountains. Far from it. We aim only to show in its true light the status of the case. "Tender-feet" have and are still striking it rich in all mineral regions. Pluck, and energy, and brains will succeed in any country. There are plenty of opportunities to make money in a new country outside of mining. Towns build up rapidly, lots double up in value,

---

* See Chapter 24th, Life of Sandy Bowers.

and merchants, grocers, tradesmen and mechanics are often suddenly enriched from small investments on the start. To all who have means and energy, and the desire to work, there is always plenty of room.

Now, to those who may be likely to go, a few hints as to what they require for the journey. First of all, those who go should seek, if possible, to go in companies of two or more trusted acquaintances or friends. It is both cheaper, pleasanter, and safer for parties to camp out in this manner by clubbing together. A common error is too much baggage. Take but little. You will need no fine clothes. If you took them, the dust and dirt of camp life would ruin them. One plain, strong suit, overalls, two or three heavy woollen shirts, and also drawers (indispensable to health with the cold nights of high altitudes); no cotton shirts, but woollen overshirts, also, and only such other clothing as is absolutely necessary, and no more than you can pack in a small satchel, all told. Go expecting to "rough it," and prepared to walk and to *carry all your traps and baggage* when in the mountains. It will be found better to buy things as you need them, than to be burdened with too much luggage. When near your destination purchase two pairs of heavy blankets, and carry them with a shawl-strap. For yourself and partner buy a small tent, a frying-pan, coffee-pot, tincups, etc. Do not buy these articles until you get into the mountains. A common error is to buy them too far away from your destination. There are always towns not far away where all such supplies are kept, and frequently the very place to which you are going is well supplied with all these articles. Such was our own experience, although we freighted our supplies from Denver, paying four cents per pound charges. We could have purchased at better advantage in the mountains.

Those who expect to prospect should, in some town as near their destination as possible, buy a burro,* which will cost about twenty-five dollars, pack upon his back a supply of provisions, an axe, pick, shovels, cooking utensils, tent and blankets. Drive the animal to any point you choose, and to transport yourselves, walk. It costs but a trifle to keep a burro; he will live chiefly upon the pasture along the way, and they are of great assistance in moving from place to place; and are remarkably sure-footed, and will climb the most rugged mountains where other animals would fail. There are thousands of them in use in the Rockies, as pack animals. Such an outfit, including the jack and tools, with a month's provisions for two persons, will probably not exceed in cost sixty dollars.

And should you, kind reader, ever take such a journey, let us hope that it may be a grand success; that you may, indeed, strike a "bonanza."

---

* The Mexican name for a donkey or jack; name used in California and Colorado.

And last, but not least, let us hope that you, like ourselves, may reach your homes again in safety, thankful to the kind Providence whose arm protected you throughout the dangers of your journey. And, like us, may you enjoy good health, and return with renewed strength and vigor of body and mind, wrought through the benign influences of the clear sky, pure atmosphere and delightful scenery of the Rocky Mountains.

APPENDIX.

VALUABLE TABLES, SHOWING THE YEARLY PRODUCT OF THE UNITED STATES FROM 1848 TO 1880—PRODUCT OF THE STATES AND TERRITORIES WEST FOR 1879—ANNUAL PRODUCT OF LEAD, SILVER AND GOLD IN THE STATES AND TERRITORIES WEST OF THE MISSOURI, FROM 1870 TO 1880—THE WORLD'S PRODUCT OF GOLD AND SILVER—THE DIVIDENDS OF THE MINES OF THE UNITED STATES FOR 1879—LIST OF THE DIVIDEND-PAYING MINES OF THE UNITED STATES—DECISION OF THE COMMISSIONER OF THE GENERAL LAND OFFICE IN REGARD TO THE TOWN-SITE OF DEADWOOD, DAKOTA—THE MINING LAWS OF THE UNITED STATES AND REGULATIONS THEREUNDER.

Valuable Tables, Showing Product of Precious Metals, etc.

Below will be found a table showing the annual product of gold and silver in the United States since the discovery of gold in California, in 1848, not including the products of lead in silver bullion:*

Year.	Gold.	Silver.	Total.
1848	$5,000,000	....	$5,000,000
1849	40,000,000	....	40,000,000
1850	50,000,000	....	50,000,000
1851	55,000,000	....	55,000,000
1852	60,000,000	200,000	60,200,000
1853	65,000,000	200,000	65,200,000
1854	60,000,000	200,000	60,200,000
1855	55,000,000	200,000	55,200,000
1856	55,000,000	200,000	55,200,000
1857	55,000,000	200,000	55,200,000
1858	50,000,000	200,000	50,200,000
1859	50,000,000	200,000	50,200,000
1860	46,000,000	1,000,000	47,000,000
1861	43,000,000	2,000,000	45,000,000
1862	39,200,000	4,500,000	43,700,000
1863	40,000,000	8,500,000	48,500,000
1864	46,100,000	11,000,000	57,100,000
Carried forward	$814,300,000	$28,600,000	$842,900,000

* Carefully compiled from John J. Valentine's Circulars and other statistics (of the Wells, Fargo & Co.'s Express Company, San Francisco).

Year.	Gold.	Silver.	Total.
Brought forward	$814,300,000	$28,600,000	$842,900,000
1865	53,225,000	11,250,000	64,475,000
1866	53,500,000	10,000,000	63,500,000
1867	51,725,000	13,500,000	65,225,000
1868	48,000,000	12,000,000	60,000,000
1869	49,500,000	13,000,000	62,500,000
1870	50,000,000	16,000,000	66,000,000
1871	43,500,000	22,000,000	65,500,000
1872	36,000,000	25,750,000	61,750,000
1873	36,000,000	35,750,000	71,750,000
1874	42,177,092	30,251,114	72,428,206
1875	42,000,000	35,703,413	77,703,413
1876	46,850,000	38,500,000	85,350,000
1877	45,100,000	38,950,000	84,050,000
1878	37,576,030	37,248,137	74,824,167
1879	31,470,262	40,032,857	71,503,119*
Total	$1,480,923,384	$408,035,521	$1,889,458,905

* Besides $4,185,769 in value of lead.

Statement of the amount of precious metals produced in the States and Territories west of the Missouri River, including British Columbia (and receipts in San Francisco by express from the west coast of Mexico), during the year 1879:

STATES AND TERRITORIES.	Gold dust and bullion by express.	Gold dust and bullion by other conveyances.	Silver bullion by express.	Ores and base bullion by freight.	Total.
California	$16,348,730	$817,436	$739,440	$285,367	$18,190,973
Nevada	163,847	......	16,622,472	5,206,395	21,997,714
Oregon	943,601	94,360	......	......	1,037,961
Washington	77,579	7,757	......	......	85,336
Idaho	1,035,804	207,160	578,336	270,000	2,091,300
Montana	1,907,053	95,352	1,194,389	432,226	3,629,020
Utah	211,640	21,164	2,559,042	2,677,033	5,468,879
Colorado	3,144,697	314,469	1,594,349	9,360,000	14,413,515
New Mexico	19,800	......	603,000	......	622,800
Arizona	212,722	21,272	1,046,036	662,373	1,942,403
Dakota	2,674,156	534,831	......	......	3,208,987
Mexico (West Coast)	92,916	......	1,249,955	341,000	1,683,871
British Columbia	976,742	......	......	......	976,742
	$27,814,287	$2,113,801	$26,187,019	$19,234,394	$75,349,501

The bullion from the Comstock Lode contains $41\frac{20}{100}$ per cent. gold and $58\frac{80}{100}$ per cent. silver. Of the so-called base bullion, from Nevada, 27 per cent. was gold, and of the whole product of the State, $27\frac{50}{100}$ per cent. was gold.

The gross yield for 1879, shown above, aggregated, is approximately as follows:

Gold, $43\frac{20}{100}$ per cent.,	$32,539,920
Silver, $50\frac{25}{100}$ per cent.,	38,623,812
Lead, $5\frac{55}{100}$ per cent.,	4,185,769
	$75,349,501

Annual products of lead, silver and gold in the States and Territories west of the Missouri River, from 1870 to 1879:

Year.	Product as per W. F. & Co.'s Statements, including amounts from British Columbia and West Coast of Mexico.	Product after deducting the amounts from British Columbia and West Coast of Mexico.	The net product of the States and Territories west of the Missouri River, exclusive of British Columbia and West Coast of Mexico, divided, is as follows:		
			LEAD.	SILVER.	GOLD.
1870..	$54,000,000	$52,150,000	$1,080,000	$17,320,000	$33,750,000
1871..	58,284,000	55,784,000	2,100,000	19,286,000	34,398,000
1872..	62,236,959	60,351,824	2,250,000	19,924,429	38,177,395
1873..	72,258,693	70,139,860	3,450,000	27,483,302	39,206,558
1874..	74,401,045	71,965,610	3,800,000	29,699,122	38,466,488
1875..	80,889,057	76,703,433	5,100,000	31,635,329	39,968,194
1876..	90,875,173	87,219,859	5,040,000	39,292,924	42,886,935
1877..	98,421,754	95,811,582	5,085,250	45,846,109	44,880,223
1878..	81,154,622	78,276,167	3,452,000	37,248,137	37,576,030
1879..	75,349,501	72,688,888	4,185,769	37,032,857	31,470,262

The exports of silver during the present year to Japan, China, India, the Straits, etc., have been as follows: From Southampton, $33,000,000; Marseilles and Venice, $5,000,000; San Francisco, $8,000,000. Total, $46,000,000; as against $39,000,000 from the same place in 1878.

JOHN J. VALENTINE,
General Superintendent.

The World's Product of Gold and Silver.*

ESTIMATE OF THE WORLD'S PRODUCT OF SILVER, LOWEST AND HIGHEST PRICE IN LONDON, IN PENCE, PER STANDARD OUNCE, AND EXPORTS FROM SOUTHAMPTON TO INDIA, CHINA, ETC., FROM 1849 TO 1876, INCLUSIVE.

YEAR.	United States.	Mexico and South America.	Russia.	Other Countries.	Total.	Lowest.	Highest	Am't exported from Southampton to India, etc.
1849	....	$30,000,000	$500,000	$10,000,000	$40,500,000	59 3/8	60 1/8	....
1850	....	30,000,000	500,000	10,000,000	40,500,000	59 1/8	61 1/2	....
1851	....	30,000,000	500,000	10,000,000	40,500,000	60	61 5/8	$8,575,000
1852	$200,000	30,000,000	500,000	10,000,000	40,700,000	59 7/8	61 7/8	12,235,000
1853	200,000	30,000,000	500,000	10,000,000	40,700,000	60 5/8	62 3/8	15,585,000
1854	200,000	30,000,000	500,000	10,000,000	40,700,000	60 7/8	61 7/8	15,475,000
1855	200,000	30,000,000	500,000	10,000,000	40,700,000	60	61 3/4	32,155,000
1856	200,000	30,000,000	500,000	10,000,000	40,700,000	60 1/2	62 1/4	60,565,000
1857	200,000	30,000,000	500,000	10,000,000	40,700,000	61	62 3/8	83,655,000
1858	200,000	30,000,000	500,000	10,000,000	40,700,000	60 3/4	61 7/8	23,765,000
1859	200,000	30,000,000	500,000	10,000,000	40,700,000	61 3/4	62 3/4	74,140,000
1860	1,000,000	30,000,000	500,000	10,000,000	41,500,000	61 1/4	62 3/8	42,390,000
1861	1,500,000	30,000,000	500,000	10,000,000	42,000,000	60 1/8	61 3/4	34,120,000
1862	3,000,000	30,000,000	500,000	10,000,000	43,500,000	61	62 1/2	50,455,000
1863	7,000,000	30,000,000	500,000	10,000,000	47,500,000	61	61 3/4	41,315,000
1864	10,000,000	30,000,000	500,000	10,000,000	50,500,000	60 5/8	62 1/2	31,270,000
1865	11,000,000	30,000,000	500,000	10,000,000	51,500,000	60 1/2	61 7/8	17,990,000
1866	10,000,000	30,000,000	500,000	10,000,000	50,500,000	60 3/8	62 1/4	11,825,000
1867	13,000,000	30,000,000	500,000	10,000,000	53,500,000	60 3/8	61 1/4	3,210,000
1868	12,000,000	27,500,000	500,000	10,000,000	50,000,000	60 1/2	61 1/8	8,175,000
1869	12,000,000	25,000,000	500,000	10,000,000	47,500,000	60	61	11,810,000
1870	16,000,000	25,000,000	500,000	10,000,000	51,500,000	60 1/4	62	7,895,000
1871	20,286,000	27,500,000	500,000	10,000,000	58,286,000	60 1/4	61	18,560,000
1872	20,527,500	26,000,000	500,000	10,000,000	57,027,500	59 1/4	61 1/8	28,270,000
1873	28,352,100	24,000,000	500,000	10,000,000	62,852,100	57 7/8	59 15/16	12,185,000
1874	30,498,000	25,000,000	500,000	10,000,000	65,998,000	57 1/4	59 1/8	35,460,000
1875	34,043,910	25,000,000	500,000	10,000,000	69,543,910	55 1/2	57 3/4	18,570,000
1876	41,506,672	25,000,000	500,000	10,000,000	77,006,672	46 1/2	58 1/2	47,500,000
Total	$273,314,182	$800,000,000	$14,000,000	$280,000,000	$1,367,314,182	....	....	$747,450,000

* From the Superintendent of Wells, Fargo & Co., Mr. John J. Valentine's Circular for 1876.

THE WORLD'S PRODUCT. 443

ESTIMATE OF THE WORLD'S PRODUCT OF GOLD FROM 1849 TO 1876, INCLUSIVE, AND OF GOLD AND SILVER COMBINED.

Year.	United States.	Australia.	Mexico and South America.	Russia.	Other Countries.	Total Gold Product.	Total Gold and Silver Product.
1849	$40,000,000	....	$5,000,000	$14,000,000	$2,500,000	$61,500,000	$102,000,000
1850	50,000,000	....	5,000,000	13,000,000	2,500,000	70,500,000	111,000,000
1851	55,000,000	$7,000,000	5,000,000	12,000,000	2,500,000	81,500,000	122,000,000
1852	60,000,000	80,000,000	5,000,000	12,000,000	2,500,000	159,500,000	200,200,000
1853	65,000,000	70,500,000	5,000,000	12,000,000	2,500,000	155,000,000	195,700,000
1854	60,000,000	47,500,000	5,000,000	12,000,000	2,500,000	127,000,000	167,700,000
1855	55,000,000	60,500,000	5,000,000	12,000,000	2,500,000	135,000,000	175,700,000
1856	55,000,000	71,500,000	5,000,000	13,500,000	2,500,000	147,500,000	188,200,000
1857	55,000,000	57,000,000	5,000,000	13,500,000	2,500,000	133,000,000	173,700,000
1858	50,000,000	53,500,000	5,000,000	13,500,000	2,500,000	124,500,000	165,200,000
1859	50,000,000	54,000,000	4,500,000	13,500,000	2,500,000	124,500,000	165,200,000
1860	46,000,000	52,500,000	4,500,000	13,500,000	2,500,000	119,000,000	160,500,000
1861	43,000,000	49,000,000	4,500,000	15,000,000	2,500,000	114,000,000	156,000,000
1862	39,000,000	46,500,000	4,500,000	15,000,000	2,500,000	107,500,000	151,000,000
1863	40,000,000	44,500,000	4,500,000	15,500,000	2,500,000	107,000,000	154,500,000
1864	46,000,000	45,500,000	4,000,000	15,000,000	2,500,000	113,000,000	163,500,000
1865	53,000,000	44,000,000	4,500,000	16,500,000	2,500,000	120,500,000	171,500,000
1866	53,500,000	44,000,000	4,000,000	17,000,000	2,500,000	121,000,000	169,500,000
1867	51,500,000	41,500,000	4,000,000	17,000,000	2,500,000	116,000,000	170,000,000
1868	48,000,000	48,500,000	3,500,000	18,000,000	2,500,000	120,000,000	168,500,000
1869	49,500,000	46,500,000	3,000,000	20,000,000	2,500,000	121,000,000	167,500,000
1870	50,000,000	38,500,000	2,500,000	22,500,000	2,500,000	116,000,000	167,184,000
1871	35,898,000	43,000,000	3,500,000	21,000,000	2,500,000	108,898,000	161,985,959
1872	39,459,459	36,500,000	3,500,000	23,000,000	2,500,000	104,959,479	170,898,023
1873	40,446,593	39,000,000	3,500,000	22,500,000	2,500,000	107,956,593	164,604,045
1874	40,102,045	29,500,000	4,000,000	22,500,000	2,500,000	98,603,045	168,789,657
1875	41,745,147	28,500,000	4,000,000	22,500,000	2,500,000	99,245,147	178,333,173
1876	44,328,501	28,000,000	4,000,000	22,500,000	2,500,000	101,328,501	
Total	$1,356,430,745	$1,267,000,000	$118,500,000	$463,000,000	$70,000,000	$3,214,930,745	$4,382,304,927

The Dividends of 1879.*

We make up the following detailed statement, showing the incorporated mines which have paid dividends during 1879:

GOLD MINES.

Name.	No.	Amount.	Name.	No.	Amount.
Amador Con., Cal.	1	$7,500	Idaho, Cal.	12	$142,600
Bodie Con., Cal.	6	300,000	Pioneer Gold M. Co., Cal.	2	8,000
Empire (Amador), Cal.	4	64,000	Plumas National, Cal.	6	54,000
Excelsior " "	9	165,000	Plumas Eureka, Cal.	1	60,937
Findley (Georgia)	4	8,000	Sierra Buttes, Cal.	1	46,875
Father De Smet, Cal.	1	30,000	Standard, Cal.	12	600,000
Green Mountain, Cal.	6	37,500			
Homestake, Dakota	12	360,000			$1,874,412

GOLD AND SILVER MINES.

Name.	No.	Amount.	Name.	No.	Amount.
Argenta, Nevada	1	$20,000	Independence, Nevada	3	$75,000
Belle Isle, "	6	300,000	Ophir, "	1	100,000
California "	5	1,620,000	Richmond Con., "	3	337,500
Con. Virginia, Nevada	5	1,350,000			
Eureka Con., "	12	725,000			$4,527,500

SILVER MINES.

Name.	No.	Amount.	Name.	No.	Amount.
Briggs, Col.	1	$8,000	Leadville, Col.	6	$120,000
Caribou, Col.	4	40,000	La Plata M. & S. Co., Col.	3	45,000
Climax, "	1	20,000	Martin White, Col.	3	60,000
Chrysolite, Col.	2	400,000	Ontario, Col.	12	600,000
Horn Silver, Col.	1	100,000	Tombstone, Arizona	1	50,000
Indian Queen, Nevada	1	12,000			
Little Pittsburg, Col.	12	1,250,000			$2,705,000

COPPER MINES.

Name.	No.	Amount.	Name.	No.	Amount.
Calumet & Hecla, Mich.	4	$1,600,000	Ore Knob, Mich.	2	$97,500
Central City, Mich.	1	80,000	Quincy, Mich.	1	40,000
Osceola, Mich.	1	60,000			
					$1,877,500

Napa Consol M. Co. ... 10,000

RECAPITULATION.

13 Gold mines	$1,874,412
8 Gold and Silver mines	4,527,500
12 Silver mines	2,705,000
5 Copper mines	1,877,500
1 Miscellaneous	10,000
	$11,094,412

Annexed will also be found a list of the dividend-paying gold and silver mines of the United States, correct, up to July 1st, 1880, showing the

---

* Compiled by the "Mining Record," New York.

amount of capital of each number of shares, par value of shares, latest quotation of stock, current value of mine, number and amount of assessments, if any; total dividends to date, and the number of dividends and amount per share. It is a very useful table for reference. It is the same as the tables published weekly by the various mining journals in New York.

Definitions of Mining Terms.

Selected from Glossary of Yale, definitions of Van Cotta, and other sources.

Adit.—A level, a horizontal drift or passage from the surface into a mine.

Alluvium.—A deposit of loose gravel between the superficial covering of vegetable mould and subjacent rock.

Apex.—The top or highest point of mineral.

Argentiferous.—Containing silver.

Ascension.—The theory that the matter filling fissures was introduced from below.

Assay.—To test ores by chemical or blow-pipe examination.

Auriferous.—Containing gold.

Bed.—A horizontal seam or deposit of mineral.

Blende.—An ore of zinc, consisting of zinc and sulphur.

Bonanza.—Fair weather; a mine is said to be *en bonanza* when it is yielding a profit.

Boulders.—Loose, rounded masses of stone.

Breast.—The face of a tunnel or drift.

Cap.—A vein is in the "cap" when it is much contracted.

Carbonates.—Soft carbonates; salts containing carbonic acid, with a base of lead. Hard carbonates; the same with iron for a base.

Cheek.—The side or wall of a vein.

Chimneys.—The richer spots in lodes as distinguished from poorer ones.

Cinnabar.—Sulphuret of mercury.

Claim.—The space of ground located and worked under the laws.

Chlorides.—A compound of chlorine and silver.

Contact.—A touching, meeting or junction of two substances, as rocks.

Contact vein.—A vein along the contact plane of, or between two dissimilar rock masses.

Country.—The ground traversed by a vein.

Country rock.—The rock masses on each side of a vein.

Course of vein.—Along its length (see *Strike*).

Crevice.—A narrow opening, resulting from a split or crack; a fissure.

Cribbing.—The timber or plank lining of a shaft; the confining of the wall-rock.

Cropping out.—The rising of layers of rock to the surface.

Cross-cut.—A level driven across the course of a vein.

Cut.—To intersect a vein; open cut, a level without a covering driven across the course of a vein.

Debris.—Fragments detached from rock or mountain.

Descension.—The theory that the material filling veins came in from above.

Diggings.—Name applied to placers being worked.

Dip.—The slope, pitch or angle which a vein makes with the plane of the horizon.

Diluvium.—A deposit of superficial sand, loam, pebbles, gravel, etc.

Ditch.—An artificial water-course dug in the earth; a flume or canal.

Drift.—A horizontal passage underground.

Dump.—A place for deposit of tailings, or waste rock.

Dike.—A wall-like mass of mineral matter filling fissures.

Exploitation.—The working of a mine; the amount of work done.

Face.—The end of a drift or tunnel.

Fault.—A displacement of strata or veins so that they are not continuous.

Feeder.—A small vein joining a larger one.

Fissure vein.—A fissure or crack in the earth's crust filled with mineral matter.

Float.—Loose rock or isolated masses of ore, or ore detached from the original formation.

Foot-wall.—The layer of rock immediately under the vein.

Forfeiture.—A failure to comply with the laws, prescribing the quantity of work.

Galena.—Lead ore; sulphur and lead.

Gangue.—The substance inclosing and accompanying the ore in a vein.

Gash vein.—A vein wide above and narrow below.

Geode.—A cavity studded with crystals or mineral matter; a rounded stone containing such a cavity.

Gold.—A reddish, yellow-colored metal.

Hanging wall.—The layer of rock or wall over a lode.

Heading.—The vein above the drift.

Headings.—In placer mining, the mass or gravel above the head of sluice.

Horse.—A mass of rock matter occurring in or between the branches of a vein.

Hydraulicing.—Washing down a placer claim by the use of hose or "giant nozzle."

Impregnation.—Metallic deposits having undefined limits and forn..

Incline drift.—An inclined passage underground.

Infiltration.—The theory that vein-filling was introduced as mineral water.

Injection.—The theory that vein-filling was introduced by an igneous fluid, and solidified.

In place.—A vein or lode inclosed on both sides by fixed and immovable rock.

Lagging.—The timber over and upon the sides of a drift.

Length.—A certain portion of the vein when taken on a horizontal line.

Level.—A horizontal passage or drift into a mine from a shaft.

Little Giant.—A jointed iron nozzle used in placer mining.

Lode.—Aggregations of mineral matter containing ores in fissures.

Matrix.—The rock, or earthy matter containing a mineral or metallic ore.

Metallurgy.—The science of the reduction of ores.

Mine.—An excavation in the earth from which mineral substances are dug.

Mill-run.—A test of a quantity of ore after reduction.

Nodule.—A rounded mass of irregular shape.

Ores.—Compounds of metals with oxygen, sulphur, arsenic, etc.

Outcrop.—That portion of a vein appearing at the surface.

Patch.—A small placer claim.

Placer.—A gravelly place where gold is found; includes all forms of mineral deposits excepting veins in place. Sec. 2329, Rev. Stat. United States!

Pocket.—A rich spot in a vein or deposit.

Prospecting.—Searching for new deposits, also preliminary explorations to test the value of lodes or placers.

Quicksilver.—Mercury, used in sluices to catch gold.

Riffle blocks.—Wooden blocks set on end in a sluice, with interstices for catching gold.

www.ingramcontent.com/pod-product-compliance
Lightning Source LLC
Chambersburg PA
CBHW032013230426
43671CB00005B/71